1·50

GW00401837

TH[

MAT[

AN[

WHELPING
OF DOGS

CAPTAIN
R. PORTMAN-GRAHAM
L.D.S., R.C.S. (Eng.)

POPULAR DOGS
London Melbourne Sydney Auckland Johannesburg

636.7082

Popular Dogs Publishing Co. Ltd

An imprint of the Hutchinson Publishing Group

17-21 Conway Street, London W1P 6JD

Hutchinson Group (Australia) Pty Ltd
30-32 Cremorne Street, Richmond South, Victoria 3121
PO Box 151, Broadway, New South Wales 2007

Hutchinson Group (NZ) Ltd
32-34 View Road, PO Box 40-086, Glenfield, Auckland 10

Hutchinson Group (SA) Pty Ltd
PO Box 337, Bergvlei 2012, South Africa

First published 1954
Reprinted 1958, 1961, 1964, 1967, 1969, 1971, 1973,
1975, 1978, 1983
© R. Portman-Graham 1954

Printed in Great Britain by The Anchor Press Ltd
and bound by Wm Brendon & Son Ltd
both of Tiptree, Essex

ISBN 0 09 152761 9

CONTENTS

PART I

THE SERVICE

CHAPTER I

CHAPTER II

PART II

WHELPING

ILLUSTRATIONS

FOREWORD

BY WINNIE BARBER

Canine Liaison Officer, The Animal Health Trust

I HAVE read in this book so much wisdom and so much sound advice that I readily commend it to all who own a kennel and especially to those who are about to start one.

The book's greatest value lies in that it sets out in simple language the difficulties that can and do crop up, first with mating and secondly with whelping.

Captain Portman Graham very wisely believes that the most intricate details are necessary to put the picture over; the infinite pains he has been to shows his wish to help not only the owner of the stud dog and brood bitch, but, and to my mind of much more importance, the dogs themselves. So much pain can be avoided by simple knowledge.

I commend, too, the fact that the author stresses the importance of calling in a veterinary surgeon in the case of the whelping bitch if there seems to be delay or any unusual symptoms.

The whole volume is made so readable as for the most part it is the result of personal experience in the author's kennel. My own experience covers Scottish Terriers, Sealyham Terriers, West Highland White Terriers, Fox Terriers and Airedale Terriers. I have been through those same scenes in all these breeds and know the value of being able to act in emergencies. This work can be equally useful to the owners of medium-sized and large breeds, as for the smaller varieties in which the experience was gained.

I think that every breeder will be the better equipped by the possession of this book.

ACKNOWLEDGMENT

MY very sincere thanks are tendered to the artist, Miss A. J. Arnott, for the excellence of her sketches, and her clever perception of the suggestions I offered. I expected diagrammatic illustrations; instead, works of art have been produced, which indicate more clearly than could any photograph the details I particularly wished to be emphasized.

Such life-like delineation combined with sympathetic artistry is the work of a genius and a true dog lover. I cannot more adequately express my gratitude.

R. P. G.

AUTHOR'S PREFACE

HAVING set out to write a book with a definite motive, it is gratifying to find that its object has been achieved.

My aim was first to invite breeders, particularly newcomers, to study carefully two important practical aspects of breeding. Secondly, I expressed confidence that breeders, by increasing their knowledge, could spare their bitches pain during a service and at whelping time; that further they could save the lives of whelps, and finally reduce the bitch mortality rate at parturition, while relieving themselves of anxiety.

Critics have expressed the view that such a book is a necessity and that its contents would be beneficial to breeders. The technical references have been endorsed by the veterinary profession, which assures readers that the biological facts and anatomical details are authentic.

I have received quite considerable private correspondence from breeders complete strangers to me, who have sent me interesting accounts of their whelping experiences. The advent of a second edition affords me the opportunity of expressing satisfaction that the book has been of some help since so many have expressed in similar terms the two following remarks which I quote verbatim: "My bitch had rather a difficult whelping being a maiden, but I followed out your instructions and am glad to say I saved every whelp." . . . "I am only a novice and this was my first litter, but I had previously read your book, sat with it in my lap during the whole evening, and everything went off quite all right exactly as you described it would, but I was thankful I had it beforehand."

When a book is reprinted the author is given the opportunity of bringing it up to date. However, the laws of nature, the facts of life, and biological mechanism are all much as they were in the Garden of Eden; they do not change, and consequently I have not felt it necessary to change the text in any way. The veterinary profession and another expert in animal breeding have however mentioned certain additional factors which I think will give added value to the subject matter, and these can with advantage be discussed here.

The first concerns the circumstance when it seems completely impossible to effect a mating. This is due to the frigidity of the bitch, and results from an anatomical structure in the vagina. This is a band of voluntary muscle which plays a big part in acceptance of the dog by the bitch, ease of penetration by the dog, and finally in the tie. It is erroneously considered by many breeders that only the dog is responsible for the tie, whereas in fact the bitch has an important rôle. When a frigid bitch will not relax this muscle the condition is often mistaken by breeders for a stricture. I suggest the only practical remedy for this difficult mating problem calls for much more patience. All hope of "getting the job over in half an hour" must be given up. The bitch may have to be kept for two or three days to become thoroughly acquainted with the dog; her apprehension and nervousness causing her to tighten the voluntary muscle to exclude the dog, must be overcome. It is the occasion when a union will only result from a "love match", since a mating will not be effected while the bitch remains frigid. With patience, she will ultimately relax; or in "doggy parlance" we must be prepared to spend a considerable time until " she warms up to her job" and voluntarily loosens the tight muscular band. If inability to effect a mating is very prolonged, it is a case where examination by a veterinary surgeon can be sought, to ascertain whether the difficulty is due to this muscular band, or whether there is in fact a definite pathological stricture or an adhesion from a previous whelping. There is a frequently confused similarity between the two conditions.

A humorous comment has been made regarding Figure 2, and I am asked how the whelp, still enclosed in foetal membrane with the placenta neatly attached, has managed to propel itself so far from the point of arrival. The answer is of course that the illustration is a line drawing, and while being exquisitely executed by the artist, it is purely diagrammatic and therefore more explanatory than any photograph could be—even if it were possible to photograph this phase at the right moment. It never occurred to me, however, that anyone would think that the whelp had walked there. I am informed that I refer to the posterior presentation as a breech, being abnormal. It certainly occurs with some frequency, but as it results so often in very slow delivery, I deemed it advisable, particularly for a novice's benefit, to describe in detail the assistance which can be given.

Another controversial point is that I have followed tradition in asserting that it is both unkind and unwise to breed from a bitch at every season. One must agree that it is generally accepted as a biological principle that the exercise of any function encourages and develops that function, and to mate regularly is considered good practice in other forms of livestock breeding. For example, the pig-breeder is always advised to have a sow mated at the first heat after weaning her piglings, which occurs within a few days, and it is conceded that a normal bitch has a rest of at least two months after leaving her litter. I maintain, however, that dogs are domesticated animals, that this book was written as advice to breeders of pedigree show dogs, and, apart from the affection they have for their charges, they also have to keep them in condition for exhibition for many months each year. In short, our dogs are not pigs. Strictly in accordance with scientific biological principles therefore, readers are now informed that the practice of mating their bitches at every season is not harmful. But if I am perhaps too dogmatic in saying this practice is cruel, I cannot retract the general advice I have given on the point in as far as it affects show dogs.

A minor comment is to the effect that this book might scare a novice contemplating breeding and perhaps deter him from his intention. I take an opposite view (now justified by letters actually received from beginners), and aver that a completely ignorant beginner can reasonably be apprehensive on occasions, while one who has been intelligent enough to gain knowledge beforehand will acquire a degree of confidence which will overcome any possibility of "jitters". If a prospective breeder is so squeamish or timid as to be put off by reading a simple explanation of a project he proposes to undertake, I suggest, in the interest of his bitch, it would be preferable for him not to breed, and instead to grow carnations. In this connexion, I repeat that the big majority of whelping cases are normal, cause no anxiety and are happy and interesting events for owner and dog, drawing them mutually closer together. The normal whelping process must, however, be understood. I have urged that an owner should know how to recognize conditions when a case is *not* going to be normal; I designed the book to give him that ability.

R. P. G.

October 1957.

I had not allowed for the advances of modern science when stating in this earlier Preface that biological processes are not subject to change and call for no revision. There is now a contraceptive pill for bitches due in season. This procedure is good, since it will prevent any accidental matings, and the frustrations resulting from a misalliance. The fact that further impressions have become necessary indicates that more and more newcomers to dog breeding have acquired a simple and fundamental knowledge, instrumental in reducing unnecessary suffering and the bitch mortality rate at parturition, while increasing the number of live whelps delivered. Since this was the purpose of the book when originally written, it is with a sense of gratification that I find its aim is being achieved.

R. P. G.

Dorridge, Warwickshire,
March 1969

GENERAL INTRODUCTION

In contemplating writing this book the author was fired by two main motives. The first, to assist the novice with two of the most intricate practical phases connected with dog breeding—(a) the mating of a dog to a bitch, commonly known as the Service, and (b) the delivery by a bitch of her puppies, commonly known as Whelping. The second motive, to save the lives of bitches and their whelps. As a result of more practical knowledge having been acquired of the whelping phenomena it is expected that the mortality rate in bitches and amongst whelps should be very considerably reduced. At the same time one hopes that the information imparted will result in relieving the bitch of unnecessary pain and suffering before, during and after the natural biological event of parturition.

The Service and Whelping are two subjects which are closely co-related, and it is pertinent that they should be studied together in one volume. Without the correct and successful performance and conclusion of the former, the latter cannot come about. Consequently it was decided to divide this book into Part I, The Service; and Part II, Whelping. This Introduction serves to explain the author's aims with the book as a whole. Part I is prefaced separately regarding the subject of the Service, and at Part II will be found an individual preface to Whelping.

In any sphere of life's activities or occupations fundamental knowledge and a wealth of detailed information of a subject will create and strengthen complete confidence in its practical application. Confidence invariably spells success, and nowhere in the various aspects of dog breeding is confidence more required than in the two phases we shall consider, particularly when an individual, single-handed, is faced with unforeseen or complicated circumstances in connection with either.

At the present time, owing to the increased and increasing popularity of dog breeding as a fascinating and profitable hobby, as a profession and as a national industry and export medium, there is an amazingly increased number of devotees and enthusiastic newcomers to dog breeding in each category. With certain aspects of this engrossing occupation newcomers are inclined to allow enthusiasm and ambition to outstrip experience. They embark on a venture of which previously they were ignorant, and for which, to ensure success, sound fundamental knowledge of details is required. It is natural, therefore, that there has been a corresponding influx of canine literature, amounting almost to a glut, from "old hands" and new authors writing on modern lines and principles. The majority of these books are excellent and invaluable to the newcomer, and their perusal will enhance his or her prospects of success both with stock and from the financial aspect. Some lack enough experience to be authoritative for novices, some lack originality to stimulate the interest of the more expert breeders. In actual fact, *Popular Dogs Publishing Company* (by whom this book is published) have produced and are producing some outstandingly good text-books on individual breeds, written by an eminent authority in each particular breed. Such books are the main source of knowledge for newcomers to that breed, with much additional advice on many other matters of dog-lore. They are also producing books on matters of general interest for *all* breeds as distinct from breed interest only. In each breed book it is inevitable that there should be a degree of similarity. In a breed book it would not be practical for a long, detailed discourse to be entered into on any one of the facets of breeding, such as rearing, housing, mating, whelping and management generally.

These subjects are usually dealt with in general terms only, to suit the breed, and though exceptionally sound throughout they are touched on only briefly. All will make a reference to whelping but none enters deeply into its intricacies, ramifications, vicissitudes or complications. Few make any reference at all to the actual conduct of the Service, yet a store of information is necessary for a novice breeder, without which he will be faced with pitfalls, disappointments, accidents and injuries (not only to the dogs, but also to himself) which, with deeper knowledge, could have been easily avoided.

With so much current literature it may reasonably be asked, "Why any more?" The author can unhesitatingly reply (in the light of his own experiences, and in view of the elementary questions he is asked regarding the Service and Whelping)—because• breed books are not sufficiently detailed; they do not tell the novice how to act for himself when necessary. Outlines need to be filled in. There is the need for a more advanced book which can act as a complement to, and amplify, the excellent though brief references to these subjects in the various breed books, and which, by diagrams, can emphasize important features. "Advanced" does not mean highly technical phraseology or scientific references likely to lead the novice out of his depth, or bore him. It means simply that he will be taken a few stages further and, by more explicit detail, be provided with an increased fund of knowledge.

In the subject matter of this book the need has been met. Of necessity some of the matter is technical, but every effort has been made to simplify abstruse medical or veterinary terms and make them easily understood by a completely inexperienced breeder.

Anyone having the temerity to write a book for dog breeders and exhibitors should realize that he has attempted a task of no mean magnitude. He has to encompass over ninety different breeds, whose type, anatomy and characteristics range from a Chihuahua to an Irish Wolfhound. As far as the two subjects dealt with in this book are concerned, there are certain general principles applicable to all breeds, particularly in connection with whelping. Regarding the Service, there are modifications according to breeds, and as far as possible these have been grouped and special methods mentioned for different breeds. When the novice has grasped fundamentals and general principles one can help him further only by advising him to watch a service or be present at a whelping of his own particular breed before attempting to control either on his own initiative. If this is under the supervision of an expert in his breed he should now be able more intelligently to appreciate verbal explanations.

Finally, any writer on a technical subject should have specialized knowledge qualifying him to do so. The author's professional work, combined with a life-long love of dogs, of which the last twenty-five years have been continuously in contact with them as a breeder, exhibitor and judge, has created a hobby of absorbing interest. The combination and culmination has been the production of this book, enabling one to pass on to others the fruits of hard experience. Its chief claim to originality is simplification of veterinary anatomical factors and biological processes, making them more easily intelligible and so ensuring confidence in their practical application. At the same time it is hoped it will inculcate in all breeders, new or old, a high esteem for the veterinary surgeon to whom their gratitude is due.

At this juncture it is essential to impress upon the reader that there is no reason for a novice to be discouraged from becoming a breeder because this book discloses that there are certain pitfalls and difficulties to be expected with the service, and teaches him how to cope with them. Similarly with whelping; the fact that sundry complications are described should not deter a keen newcomer from proceeding with his plans. In all normal circumstances a whelping case is a particularly fascinating, interesting and happy event for the owner, and a natural one for the bitch. But don't let us be ostriches and blind ourselves to the fact that whelping cases are not *always* normal or simple. One will not go so far as to say that the management of a service can be as pleasurable as a normal whelping but it certainly is not devoid of interest; and with knowledge many services are managed with the utmost ease. But it would be foolish to delude a novice into imagining that he will never experience trying and untoward situations.

The possibility of a newcomer being deterred from taking up dog breeding as a result of the detailed information given on these two subjects has been considered. Actually, several eminent breeders of various breeds were approached as to the advisability of imparting such information. The general opinion was that it would be beneficial. The author holds the view that some newcomers tend to drop out of the game, and thus are a loss to dogdom, if at a very early stage in their breeding careers they are completely discouraged by some unforeseen occurrence connected either with the Service or with Whelping. Had such beginners had some pre-knowledge of the subject, the contingency which deterred them would have been anticipated and not have taken them unawares. Confidence, based on information gleaned from prior reading, would quite possibly have *prevented* the happenings which put them off. At least it would have reduced their significance and enabled the beginners to tackle their difficulties. They would be far more likely to continue with their hobby when in a position to surmount an early snag instead of "falling at the first hurdle" because of ignorance. Cure of such ignorance is the purpose of this book.

Controversy is more stimulating than dogma, since

it creates an added interest. It will be for the novice reader himself, and the dog-breeding public in general, to draw their own conclusions, remembering that it is not only the interests of owners which have been considered but also the welfare of the bitch, her progeny and the stud dog.

R. P. G.

Hampton-in-Arden.
1954.

PART I

The Service

PREFACE

I T IS a common lay fallacy to consider that all that is necessary for a dog to serve a bitch, and to obtain a litter of puppies, is just to put them together somewhere and expect that a mating will automatically take place. Nothing could be further from the actual facts. The majority of matings are straightforward, some are difficult, a few are extremely difficult, and in rare cases failure and complete inability to effect a mating will result. Experienced breeders with memories of some exceptionally arduous services will readily agree that the novice owner of a stud dog has a great deal to learn. He will find in the following pages all the reasons why the service cannot always be the simple, automatic act he may have thought it was. He will learn how complications and setbacks can arise, and, more importantly, he will be instructed in various methods by which difficulties can be overcome, and told of measures whereby he can prevent injury to the dog or the bitch or possibly to himself.

Let us imagine an example of a sequel to "novice's luck". There have been several instances of the latter in almost all breeds during the post-war years. (Incidentally, such instances do tend to quash that erroneous, unsporting impression that only experts win; many recent examples have proved that if a novice has a really good dog he gets an equal chance in competition with experienced exhibitors.) A man decides he would like to have a dog. He gets a dog pup and a knowledgeable person tells him it is a top-ranking specimen and he ought to show it. Thrilled at the prospect, he does so with marked success. He is getting caught in the vortex of the "dog game" and his increasing zeal takes him to his first Championship Show. Very

23

shortly he wins a Challenge Certificate, then another, then another, and with a dizzy head he finds he owns a Champion Dog. He had previously never seen a service and the extent of his knowledge of stud work was that he knew some dogs are dogs—that is, males—and some are bitches—that is, females. While his dog's rapid rise to eminence was in progress shrewd heads had made the mental note: "When such-and-such bitch comes into season next I'll use that dog for her." The result: one day he receives a telegram telling him that a bitch will shortly visit his dog. Earlier, and before getting his "third", he may have rashly advertised his dog in glowing terms and at a large fee in the jargon he had observed in the canine Press. Puffed with justifiable pride at the honour done him by receiving such a valuable bitch from a Kennel with such a reputation, he collects the bitch from his station. Nonchalantly he opens the hamper on the lawn, a bitch leaps out and (the gate of course being left open) streaks down the lane. After a cross-country run and hue-and-cry on foot, or by exceeding the speed limit in his car through the streets of a town, he is eventually lucky enough to retrieve the bitch. She promptly, and quite rightly, bites him. This is only his first bite, and only a few drops of sweat have so far been shed. More are to come. It never occurred to him to take a collar and lead, but someone who helped to catch the bitch lent him some rope with which this valuable bitch was reluctantly and in undignified manner dragged back. Home again, and this time quite pointedly securing the gate, the service barometer starts to rise. With "do or die", "now we will get on with the job at once", expressions he drags an extremely frightened, very snappy and unhappy bitch into a shed. This being cluttered up with garden para-phernalia and a collection of obsolete household utensils, but devoid of a table or a chair or a strip of matting, straw or any sacking, it was inevitable that the bitch, on release from her rope lasso, should dash away and get entangled with the lawnmower in her endeavour to hide behind something. Note that it never struck our enthusiast to realize that the bitch had had a long and frightening train journey, and might be thirsty and hungry, and would probably be very scared after finding herself running in a panic in strange surroundings. These discomforts would so easily have been allayed by the humane thought of postponing an attempt at operations for two or three hours, at once putting the bitch in a warm bed of straw or wood wool with a hot drink and food by her. Having observed her highly strung state, the addition of half an aspirin tablet or a quarter phenobarbitone to the drink would have settled her down. Again, it was an oversight not to have ascertained her pet or Kennel name, use of which might have induced a more friendly approach; instead, and being unaccustomed to being addressed as "Come here, you bitch!", she naturally would try to jam herself behind the garden roller for added seclusion, and snap at any attempt to dislodge her. "Perhaps she will come out if I bring the dog in," thought the stud manager (what a hope!) as he promptly went to fetch the latter, with a self-satisfied expression of dawning intelligence as he remembered to secure the door as he left. The young dog, with no sexual experience whatsoever (being the sole dog owned), when brought in (loose) to the shed sniffs appreciatively and in some uncanny way senses something entirely new has occurred in his short, sweet life and, from his expression, something that promises to be interesting and full of fun. He espies the bitch (now trying to get still farther behind the roller) and as it dawns on him that she is responsible for all his excitement, what more natural than for him to run quickly, playfully and joyfully to greet her on the other side? For this boldness he receives a really nasty bite on the end of his nose. He would have received many more elsewhere, had he not retired as quickly as he made his advances. Sudden ardour is diminished; what seemed likely to be enjoyable didn't seem to be so funny after all, and more attention is being paid to his nose end, now bleeding copiously, as he mournfully paws it. (Of course there was nothing at hand to staunch it.) The stud manager, a little more nonplussed, decides that some sort of next move is obviously up to him; he must forcibly drag the bitch from her hiding-place and try firmness to induce her to begin the performance of the object of her visit. Striding determinedly towards her, and treading on a rake whose handle gave him a whack on the forehead, he makes a grab at the bitch's scruff, and receives another painful bite. To have put on a pair of thick leather gardening gloves was quite beyond his perception. Though decrying his lack of preparedness, one can but admire his dogged determination and resourcefulness, born of "unnecessary" necessity. His next move was to throw an old rope potato-net over her head with one hand and forcibly pull her out by the tail with his other; the dog meanwhile being too peeved over his "bloody" nose to take much further interest in the proceedings.

After a struggle and one or two more minor bites, the rope net was wound round the bitch's head to serve as a crude muzzle. Thereafter all three parties seemed a little happier; the owner felt more secure, the bitch became more docile and the dog, realizing that he had nothing more to fear from the head end, renewed his investigations of the other. Whereupon the bitch promptly sat down, and any attempt to make her stand on her hind legs merely induced her to roll over on her back. It was fortunate that her intended mate was only a shy young dog; a sour old veteran stud has been known to bite a bitch for this perverseness. Our youngster merely sat back looking very puzzled and sheepish. The owner's grim determination was not exhausted; perhaps straddling her and gripping her between the knees would keep her upright? She resignedly succumbed to such force. Meanwhile, the dog was several yards away and now would not act on his own initiative. By dint of much patient coaxing he did approach, and a tentative mounting was essayed. But the owner was so inexperienced that it only dawned on him slowly that perhaps it would have been advisable to have had assistance from another person, that is, one to attend to the dog and one to attend to the bitch. After alternately helping dog and bitch, but finding that on releasing the one the other did everything he did not want it to do, the owner's dilemma reached its climax. His exertions from the time he opened the travelling box had been continuous for two-and-a-half hours, and he was just about where he started. He did not want the bitch's owner to have the extreme disappointment of losing "the season" by no mating, he did not want to lose his first fat fee, and he did not want to be a subject for ridicule. Consequently, with increasing perspiration from both physical and psychical turmoil, and somewhat bloodstained, by his own blood and that of both dog and bitch, he now resorts to the telephone, requesting advice from an experienced breeder who, fortunately, was not too distant from him. To his unbounded relief the breeder, as is typical of the *camaraderie* in the "dog game", replied that he would come over right away. On hearing the sad story, he took a quick glance at the bitch's vulva, and his instructions were crisp. "She'll hold all right till tomorrow. Make her a warm bed, leave a hot drink and a good meal beside her. Here is half a grain of phenobarbitone to put in her drink. Keep her quiet and keep the dog as far away from her sleeping-place as possible. Have a stiff whisky-and-soda yourself. I'll be along at nine in the morning and fix things up for you."

And so a disillusioned young stud dog went back to his kennel, thinking that this was just not his idea of love at first sight, and deciding that after all he preferred his bones.

Many of the various ways in which the experienced

breeder could and did "fix things up" the next morning
will be gleaned from the ensuing chapters. The owner
retired that night not feeling too happy, but the next
night, after things had been satisfactorily fixed up, he was
a less sad though very much wiser man. One could
go on elaborating indefinitely, but we will draw a veil
over further quandaries which might have occurred had
our novice, single-handed, reached the stage of an actual
union. The foregoing facetious episode is not, however,
without a basis of truth. One has heard of practically all
the situations at some time or another, from novices who
have sought assistance when unable to achieve the desired
result on attempting the first service for their dog. The
narrative was concocted to show also that an earlier
statement—that the owner of a stud dog has a lot to
learn—was not an exaggeration. This will be appreciated
if the reader will refer back and count the number of
actions which should, or should not, have taken place.
He will find that even in the few pages of this preface no
fewer than twenty tips and wrinkles have been given.
Knowledge of these would either have prevented the
débâcle or certainly would have eased many of the
situations.

When the reader has finished this book he will find
that the number of similar pieces of advice regarding the
service will have reached considerably more than three
figures!

Why is it that, among all the canine literature available, there does not appear to be any efficient, detailed
instructions on the actual act of sexual union, or a
description of the anatomy concerned, or an outline of the
biological and physiological phenomenon which takes
place? Without this fundamental knowledge it is obvious
that no breeder, particularly a novice, can manage a
service to the best advantage or effectively meet and
overcome the various difficulties which can occur.

The writer has come to the conclusion that the lack

of instruction on this subject is due to its delicate nature
—that is, to a form of mock modesty. He considers such
a view foolish; we live in an enlightened age, and the
Victorian era has long passed. If a man or woman takes
up dog breeding as a hobby or a profession, the management of the service is an important part of that occupation;
details should be studied and knowledge gained in the
same way that intimate knowledge has to be acquired
in the nursing, medical or veterinary professions. Consequently he has decided to get down to brass tacks and
facts, and in Chapter I will be found information considered essential. The matter is discussed with discretion,
but no detail has been omitted, either in that chapter or
subsequent ones, if inclusion was deemed likely to be
valuable. Many of the details are mentioned in order
that the prospects of a young and valuable stud dog
shall not be impaired by mishandling during its earlier
experiences. Many details will concern the bitch, to ensure
her comfort and diminish any possibility of pain at the
time of her union.

Connected with the carrying out of a service there are
certain responsibilities of the stud dog owner to the
owner of the bitch, and *vice versa*. Attention to small
details will enable arrangements to be conducted smoothly
and brought to a harmonious conclusion for both parties.
These matters are discussed in Chapter II.

As to the management of the actual union of the dog
and bitch, the first thing to be learnt is the impossibility
of expecting hard and fast rules to be laid down, or to
expect that every service is managed in the same way. It
is for this very reason that some treatise on the subject is
so necessary. Stud dogs have their own individual
techniques, and these can be classified according to their
individual behaviour. Equally, perhaps more, numerous
are the peculiarities of bitches, which vary according to
circumstances. In Chapter III an attempt has been made
to review the contingencies which may result from the

mannerisms of individual dogs and bitches at mating time.

Again, there is no cut and dried METHOD for managing any service; variations are adopted and adapted entirely according to circumstances. Some of these are due to the idiosyncrasies dealt with in Chapter III. Many are due to the position of the owner, that is, whether he must manage alone or can rely on a second person to assist. Thus, methods can be divided into: the natural or love mating; the single-handed assisted mating; the assisted mating by two persons; and specialized methods according to breeds, such as the table mating, when very frequently manipulative assistance is given, which can be termed the "guided" mating. Occasionally when the last method becomes imperative it is termed a "forcible" mating. In Chapter IV each method is explained in detail.

Chapter V deals with unwritten laws regarding the financial relations between the stud dog owner and the owner of the bitch; they can be termed "recognized customs and usages", and some affect the reputation of the owner of the stud dog for straight dealing. Their observance will enhance his chances of success as a breeder, and his Kennel reputation. Many old-fashioned proverbs are sound even though hackneyed. An example is, that in all life's ramifications honesty is ultimately the best policy; applied to dogdom there could be no greater truth. With many dog transactions, particularly those including stud work, pedigrees or the actual result of a service, it is easy to cheat. Reflect, however, that it is equally easy for the other party to realize that he has been "done" and, smarting under an injustice, he will spread the news like a prairie fire. Any trivial, immediate gain by some petty deception, or act of actual dishonesty, is immeasurably outweighed by subsequent loss, since it calls up another old truism, "Once bitten, twice shy." An owner, or a Kennel, with an impeccable reputation not only gains respect from novice and expert alike, which no one can shake, but is also the one whose stud dog, or dogs, will be recommended, and whose stock is in greatest demand. Straightness ensures success; "fiddling" means failure.

The importance of keeping complete records of stud work is underlined in Chapter VI, where the nature of such records is explained.

A preface is usually brief. This one has intentionally been long since it might have been argued that nothing not already known could be written about the service. If, however, in this *résumé* either the novice or the expert has found reference to some fact new to him, then perusal of the book will be worth while.

CHAPTER I

Simple explanation of the anatomy of the reproductive organs of dog and bitch—Biological processes of mating—Period of ovulation; signs and symptoms of ovulation; its duration—Period of readiness for mating—Some general considerations: (*a*) Regarding the bitch: when a bitch can be first mated; how frequently she may be mated; when mating should cease. (*b*) Regarding the dog: when a dog should first be trained to stud work; extent to which a dog should be used; age up to which a dog may be used—Practical application of anatomical knowledge.

SUBSEQUENT pages will indicate that it is always necessary for a human to be present at the mating of a pedigree dog to a pedigree bitch. Further, it is certainly highly desirable in some circumstances that actual assistance be given at the service. No breeder can adequately render such assistance unless he has a fundamental knowledge of the anatomy, construction and "mechanism" of the reproductive organs of each sex. Similarly, in Part II (Whelping) the anatomical situation of the bitch's generative organs will assist an owner in many aspects of this phenomenon, particularly if intervention in an emergency proves imperative. In both cases the owner with a pre-knowledge of this subject can more easily understand expressions used by a veterinary surgeon.

THE DOG.

THE TESTICLES contained in the SCROTUM are two in number (see note, page 34). These manufacture or secrete the seminal fluid containing innumerable spermatozoa which are the sperm cells or male seed. This fluid is conducted *via* a cord to the actual male organ, the PENIS, in the anterior part of which is a bone, varying in size from half an inch to four inches according to breed. In all ordinary circumstances the organ is protected by a sheath

32

of skin, technically called the prepuce, loosely attached at the bulb and front end. The only parts of the penis to which a stud dog owner's attention should be drawn are: the front part, composed of erectile tissue ending in a fine pointed end; and behind this a rounded bulb also composed of erectile tissue. It is due to this latter structure that in mating the phase known as "the tie" is brought about. This is of importance to the stud manager. From the bulb two large veins pass on to the back of the penis, where also are situated two muscles near the bulb. The purpose of this "mechanism" is threefold. Firstly, sexual excitement brings into operation the erectile nature of the two parts of the penis, thus enabling them to negotiate and penetrate the female passage. Secondly, sensory frictional nerve impulses induce the ejection of the fluid at the extreme end of this passage. Thirdly, the muscles referred to strongly and simultaneously compress the veins mentioned, causing them to become engorged in the bulb. The bulb when inside the female passage becomes so extremely congested and swollen as to cause the external structure of the female passage to be completely "locked" round it. This swollen condition of the bulb may persist for periods ranging from two minutes to one hour. The dimensions of the swollen bulb are remarkable. A comparison which may help the complete novice to visualize it is to compare it with a medium-sized, hard tomato, in small breeds up to 12 lb. There is an increase in size in proportion to the size of breeds.

Mention has been made of possible injury to a dog or a bitch. THIS LARGELY OCCURS OWING TO THE TIE, for which reason a detailed description has been given.

It should further be noted that the organ remains entirely flexible. While in the position described above, and ejection is taking place, the dog can and should be faced in the entirely opposite direction. The reasons for this will be described in later chapters. The method is colloquially known as "turning" the dog.

NOTE: When only one testicle has, at puberty, passed down into the scrotum, or is visible, the condition is known as "monorchidism", and the dog is referred to as a "monorchid". There is difference of opinion as to whether a monorchid should be bred from. There is evidence that the condition can be transmitted, that some monorchids will be in the progeny of a monorchid. At the same time there is little if any evidence to show that a monorchid is incapable of producing stock, or is non-fertile. On the contrary, there are quite a large number of examples, in all breeds, of monorchids being extremely prolific and pre-potent sires. On the Continent, and in some other parts, monorchidism is regarded as a fault. Judges are instructed to look for this condition, and if it is observed the dog is either penalized or disqualified. Under existing Kennel Club rules in this country no such instruction is given, nor is the disqualification of a monorchid considered. On this controversy the writer will not venture a definite opinion, except to say that it would appear to be quite illogical to disqualify a dog for a condition that has no deleterious effect on the species, and does not prevent the dog from being as prolific a sire as a normal dog with two testicles visible in the scrotum.

There is a further condition known as cryptorchidism, in which neither testicle has descended to the scrotum. The retained testicle (or in the case of the cryptorchid both retained testicles) is prone to atrophy or shrink and wither away, and ceases to be capable of producing seminal fluid and spermatozoa. Thus in the case of a monorchid with one testicle functioning normally the lack of the other is immaterial, but in a cryptorchid the ability to reproduce would be extremely doubtful and the specimen could not be regarded as a sire in all general circumstances, even if an exceptional case disproved this view. A strain in which cryptorchids have appeared will continue from time to time to maintain this factor of further cryptorchids periodically cropping up. The writer has not the authority to make a positive statement on the subject in a book of this nature; it is touched upon only in order that a novice may know the meaning of the terms. But it would appear to be logical to say that a monorchid in a Kennel or strain can be regarded as having no detrimental effect on the strain, whereas cryptorchidism should be avoided and eliminated.

THE BITCH.

The genitals of the bitch are (a) External, the VULVA, which is a thickened, fleshy orifice forming a pointed end facing downwards. Colloquially it is referred to as the "lips" of the vagina. (b) Internal, (1) the VAGINA, or female passage, which is relatively long and narrows as it progresses forward into the body. It houses a structure composed of erectile tissue, corresponding to that in the dog, and known as the "glans clitoridis", which provokes for the bitch the acceptance of the dog. The vaginal passage ends in a constriction as it merges into (2) the UTERUS, or womb, at the cervix or neck of this organ. The uterus has a peculiar formation, in that its "body" is comparatively short—about one-third of the length of the vagina. The main portion of the canine uterus is constituted by the body dividing in a V shape, into two horns which go forward on each side in the abdomen for a considerable length. The uterine horns are practically straight, narrow in diameter and relatively long, being approximately six times the length of the body of the uterus.

Each uterine horn ends in a structure known as (3) the OVARY, which is only slightly smaller in size than the body of the uterus. Proceeding from each ovary is a tube known as (4) the FALLOPIAN TUBE, which ends approximately half-way down each horn.

Such, then, is the anatomy of the generative system.

Anatomy is a description of structures and organs; physiology is a description of the function of those organs. Very briefly, the physiological function of the generative system is that the ovaries produce ova, or eggs, which pass down each fallopian tube and thus gain entrance into each horn of the womb. This transference of the ova from the ovaries, *via* the fallopian tubes, into the uterus takes place only at definite, specific periods, the cycle or phase having a variety of names, the technical ones being—the period of ovulation, the oestrum, or the period of oestruation. The condition in doggy parlance is described as "in season", "in use" or "on heat". The normal period intervening between each season is six months. The biological process of ovulation will be described in detail later. Provided the ova of the bitch and the sperm cells, or seed, of the dog are *both* fertile (see note below) at the time of the union a variable number of ova are "pierced" by innumerable sperms, only one of which, however, is selected. The cells of an ovum and the cells of a sperm then undergo fission or "dividing up", the whole process being known as the fertilization of the ovum. Those ova which have become fertilized attach themselves very firmly to the membranous walls of each horn of the womb. Many ova not impregnated by semen pass out as a mucous discharge subsequently. During the ensuing weeks, known as the period of gestation, further development from the two original sex cells goes on apace, their fusion forming an embryo or foetus. The period of gestation normally lasts 63 days, at the end of which the foetus is a fully developed whelp, waiting to see the light of day. Such, then, was the beginning and making of our future C.C. winner.

NOTE: Thanks to modern science and the advanced research work of the veterinary profession, the owner of an exceptionally fine specimen of a bitch or a dog, perhaps a Champion, need no longer suffer the disappointment of finding either that she is a "non-breeder",

i.e. non-fertile, or that he is impotent as a sire. Similarly
with bitches who either fail to come into season at all,
or ovulate very irregularly, going for very long periods
beyond the normal six months. These disappointing in-
adequacies can be rectified. Hormones can now be injected to bring about ovulation; another type of hormone
injection will increase the number of ova as also their
fertility. Thus a litter can now be made possible and its
size increased. Again, with a dog which has been used
at stud, say on four occasions, and at each no litter has
resulted from different bitches of proved fertility, or
perhaps only one puppy, it is possible for the spermatozoa
to be examined microscopically and any inadequacy on
the dog's part detected. In accordance with the findings,
suitable hormone injection can be administered with
beneficial result. This being a purely veterinary subject,
it is not one within the province of this book, though it
appeared advisable that the reader should be told that
by recourse to veterinary treatment such disappointments
can generally be avoided. In the course of the last eight
years the author has had actual experience with each
type, and can speak with conviction of the efficacy of the
treatment.

PERIOD OF OVULATION.
    The occasion on which a bitch first comes "into
season" is variable. One bitch may begin very early at
seven months of age, another may show no indication of
sexual change until well over twelve months old. Among
almost all the breeds of Eastern origin, e.g. Pekingese,
Chows, Afghans, etc., maturity comes very much earlier,
bitches of these breeds often starting "in season" at six
months. As a normal average, however, the first oestrum
is apparent at about ten months. Thereafter in a normal
bitch it recurs with precise regularity every six months.
In the periods between the seasons there is no possibility
of a bitch being mated; and unless a pervert (rare in

adults), neither the dog nor the bitch has any sexual interest for the other, and both can live together in harmony, unhampered by any sex urge and with no embarrassment for an owner.

There are, however, variations from the normal due to some glandular condition, when some bitches either fail to come into season at all or, after an initial oestrum, will cease to do so again, or be extremely irregular. Conversely, some bitches are known to come into use much more frequently than the routine six-monthly period. (See note on pp. 36.)

SIGNS AND SYMPTOMS OF OVULATION.

These are unmistakable and easily recognizable. The very earliest is an enlargement in the size of the vulva, and the first signs of a tendency for a dog to become sexually attracted to the bitch. It is an important factor for an owner to observe as it is an indication that the oestrum is imminent though not actually begun. Another sign not generally known, but definitely noted with bitches kept indoors, is that for at least a fortnight before the oestrum begins there is marked frequency in urination. An otherwise clean house bitch will constantly be falling from grace. The same condition occurs with Kennel bitches, but is not so likely to be seen by the owner. This changed condition of the vulva may persist for a day only or for several before the next definite sign appears. During this time the vulva changes from a deep red to purple colour and is hard and swollen. Following this, and at variable periods, the next sign is a discharge, rapidly turning to a very dark red blood colour and rather thick. The day this discharge occurs is of the utmost importance to the owner and the date must at once be noted. It is from this sign that the date on which the actual mating shall take place is decided. The expression is that the bitch is "showing colour". The period of ovulation, that is, the duration of "the heat", may last anything from nine to twenty-one days, and in exceptional cases for periods of over a month; it is variable in individual bitches and it is important for an owner to make a record of the habits in this respect of each of his bitches. Normally, after the first week from the first "show" the vulva, although remaining very enlarged, becomes looser and the tissues much softer, while at the same time the colour of the discharge lightens and by the tenth day is usually only a faint pink, gradually disappearing and generally colourless by the fourteenth day. Another most important thing for the owner to note is that the bitch is not mateable, nor desirous of being mated, throughout the whole of this period. The term "being ready" is applied to the time within this period when she will be willing to consort with and invite intercourse from the dog, indicating this by "curling her tail" from the cover it normally gives to the vulva, and quite obviously "standing", and frequently thrusting her hind part towards any dog near her. During the whole period of oestruation a bitch causes sexual stimulation in any dog, and his desire to mate is persistent throughout, though, as has been already mentioned, this is not so with the bitch. This stimulation is induced because the discharge from the bitch has an individual smell, to which the dog is very receptive, and reacts naturally and instantly. This smell cannot be detected by human beings, so causes them no unpleasantness. It is essential, therefore, that the "in-season" bitch must be kept closely confined, preferably under lock and key, for a period of at least three weeks. It is unwise to keep an "in-season" bitch in the house if a dog also is kept indoors; the dog would be under a persistent strain likely to harm his nervous system. It is kinder to the dog for the bitch to be segregated in an outside kennel. It should have a small run, but should be completely enclosed either by boards or netting for a height of six feet, as it is remarkable how acrobatic dogs can become at this time. With the bigger breeds, for additional precaution the run should also be provided with a close-fitting roof. A bitch thus "in purdah" must be exercised; it is not only cruel, but also detrimental to her well-being, for her to be cut off from normal existence for the whole of the period.

To prevent the nuisance of other dogs in the neighbourhood getting "scent" and visiting the kennel, one can either carry the bitch for 200 yards or take it a short distance in a car, and then walk it, on a lead, in a field or other suitable place. The "scent" is then not followed to the kennel. Various proprietary preparations are obtainable which, if applied to the bitch's rear legs, mask the odour and deter dogs from approaching. They are quite useful when out exercising the bitch but could not be relied on to prevent a dog ultimately being attracted and effecting a mating if both were left together for any time. Despite precautions having been taken, it occasionally happens that a bitch will get mated quite unintentionally, either to a stray dog or to a kennel inmate not intended as her sire. Such an unfortunate occurrence can be rectified if a veterinary surgeon is consulted immediately afterwards. An injection can be given which will abort a possible conception. There must be a popular belief, since the writer has so frequently been asked the question, that if a pedigree bitch has made an alliance with a mongrel, or dog of entirely different breed, she is ruined, and is of no further use as a breeding bitch. Such an idea is of course entirely erroneous. The resultant litter would be valueless as pedigree stock but, provided the whelps were of a size commensurate with the normal ones of the bitch's breed and did her no harm at parturition, she would at subsequent seasons be perfectly normal in all respects when mated to a pedigree dog of her own breed.

Having referred to minor factors concerning the phase when a bitch is on heat, we can now revert to the main subject. Another quite variable factor is the actual period at which a bitch is described as being "ready" to be mated. The behaviour of the bitch at this stage of "readiness" has been mentioned, but duration is by no means constant with all bitches. In some cases it is only a matter of a few hours when a bitch will stand and be willing for a dog to mate her. Others will stand readily to the dog at any time during periods varying from two to six days. In exceptional cases some bitches are ready for a whole fortnight. An owner should always keep a record of the behaviour of each of his brood bitches in this respect. As a general rule it is rare for a bitch to be mate-able before the ninth day, or after the fourteenth day, though extreme cases are known when a mating has been effected at the eighteenth day, with a litter resulting. In the majority of cases the eleventh day is found to be the most suitable for a bitch to be receptive to a mating. From the foregoing it will be appreciated that it is advisable to observe the bitch frequently during her season, particularly when an arrangement has been made with the owner of a stud dog for the service. If the discharge ceases early, and there are indications of the vulva shrinking and becoming smaller before the expected date (the condition being colloquially described as "drying-up"), it will be necessary to make fresh arrangements and hasten the despatch of the bitch. This avoids the disappointment of no litter resulting, because the bitch had "dried up" before a mating could take place.

It should be emphasized that although a bitch may have been satisfactorily mated at the eleventh day she may remain vulnerable for several days afterwards. All precautions regarding her seclusion to prevent an indiscretion on her part must be rigidly maintained for the full twenty-one days for safety, and not ended until every trace of the oestrum has cleared. Usually, she will then herself "see every dog off" by snapping at it in no uncertain manner, and such a surly reception usually decides the dog to give up the contest.

Some General Considerations.

(a) *Regarding the bitch.* An aspect of breeding which may fittingly be considered in this chapter is when a bitch should be first mated and how often she should be mated. Here one can almost lay down a general rule, that if a bitch comes into season at any age before she is twelve months old it is unwise to mate her at her first season and preferable to wait till the next season, when she will be more matured in every way, physically and temperamentally. An exception is made for Miniature breeds, where whelping is generally a little more difficult than for bigger breeds, and a bitch may be mated at her first season if not earlier than nine months old. If a bitch is very late in starting her first season, and does not do so until between twelve and eighteen months of age, it is in this case in order to mate her at her first season. A bitch should not be expected to deliver puppies like a machine every time she comes into season; such a procedure is cruel and heartless and not really lucrative since both the bitch and the quality of her whelps soon deteriorate, and in the long run the practice results in loss. Owners with great love and respect for their bitches will mate them only every other season, approximately a litter a year, resting one season and mating the next. An intermediate stage, whereby a strong, healthy bitch does not suffer any deleterious effect, is to mate her twice only in three seasons; that is, if she has been allowed to have litters at two successive seasons, invariably she should be rested for the next. If a bitch has been a maiden for three or four years there is considerable risk of her having a very difficult time at whelping, even if she conceives. A bitch who has had many previous litters, and proved an easy whelper, is safe to go on mating up to the age of seven or eight years. Beyond that age it is regarded as both unwise and unkind, though one has heard of bitches producing litters comfortably when twelve years old; usually, however, the progeny degenerates.

(b) *Regarding the dog*. We will now consider the ages and extent of the use of stud dogs within wise limits. Firstly, however, it must be impressed on the novice stud dog owner that his dog must be *trained* to his peculiar profession of combining business with pleasure. A dog who does not gain his first experience while young is usually very difficult later to make into an intelligent stud dog. A dog who has not been put to a bitch until he is, say, three to four years old will most probably have no idea of what is expected from him and fail to effect a service. The best age at which to start training a stud dog, and at which he should be introduced to his first bitch for a service, is ten months. He should not serve again for another two months, and another two months should elapse between the next service, when he should be allowed to serve not more than one bitch a month until eighteen months old. These are given as maximum intervening periods; with a robust, virile dog, after fifteen months old an occasional additional bitch would not be really detrimental, but to over-use a young dog under eighteen months is to weaken him, and if carried to excess, to ruin him, however much he is in demand or the magnitude of the fees he could earn. There is a pre-war case on record in a certain breed of an outstanding young dog whose value ran into four figures and who earned in his first year an amount not far short of four figures. To do this, however, while still so young, he was being grossly over-used, serving bitches on two, three or four successive days. He became completely worn out, and was dead before he was two years of age. A little mental arithmetic will quickly show the folly and shortsightedness of such financial greed. Had he served a third of the number of bitches in the first year, and half that number in the second, he would have earned only approximately £400, but in all probability would have lived and served and earned an increased amount for a further six years, representing at least £3,000. Thus in the long run the cruel abuse of this dog represented a loss of some £2,500 to the owner. After eighteen months of age a dog should not serve a bitch more often than once a week on an average. The fact that a dog is capable of, and willing for, more frequent services does not alter the fact that it is unsound to abuse the animal. To overstep this rule occasionally would not necessarily be detrimental, but to exceed it regularly over a long period would be unwise, if the owner hopes to make the dog a lasting sire of sound, robust stock. The starting age of ten months need not be slavishly adhered to, as in special circumstances a knowledgeable dog can be used younger; ten months is quoted as the optimum age, nine or eight months are possible, younger is unwise and a result unlikely. We were, however, ourselves guilty in very special circumstances of deviating from the above advice quite recently. We had a promising dog pup only just turned seven months, when due to fulfil a commission and depart to Australia. We were particularly anxious to retain his blood line before he left us. Candidly, we never considered a result likely, but we thought we would just try him at stud on the merest offchance. To our surprise he functioned perfectly; the bitch proved in whelp and in due course delivered six puppies. We have heard that he was used in Australia about three months later (aged ten months) and another sizeable litter resulted. Apparently such an early service was not detrimental. It should be noted that the bitch to whom this exceptionally gay Lothario was mated was a proved old brood of six years. An axiom is: to an old brood bitch use a quite young dog; and let an old or ageing stud dog serve only quite young or maiden bitches. The result in progeny is generally a larger-size litter with no degeneracy. To mate a very old dog to a very old bitch is not always sound; the size of the litter is invariably diminished, and deterioration in progeny usually results. Frequently only one puppy arrives, of exceptional size and sometimes monstrous and dangerous to a bitch at whelping time. The age up to which a dog can profitably be used at stud, with prospects of the bitch owner having an honest deal and a fair prospect of a worth-while litter, varies with individual dogs and quite definitely in individual breeds. No rule can be laid down. As a guide, it can be said that by ten years of age a stud dog has aged; if he has been used extensively and regularly over the years he would, in regard to sexual competence, be older than a dog of the same age who had been used much less. Approaching or beyond ten years, the activity of the sperms diminishes and smaller and degenerate stock can be expected. Dogs have been used at twelve years, and in extremely exceptional cases later, but a dog of twelve years is equivalent to an old gentleman of eighty-four years, and though the spirit might be willing the flesh will undoubtedly have weakened. In spite of the fact that many a good tune is played on an old fiddle there are perhaps limits.

PRACTICAL APPLICATION OF ANATOMICAL KNOWLEDGE.
The sole reason for describing in detail anatomical structures at the beginning of this chapter was to show the value of such knowledge when applied practically either to the supervision of a service or to various phases connected with whelping. In Part II the latter is dealt with; this chapter is concluded by advice on the practical application to a service.

With a big majority of breeders it is customary to note:

1. THE ANGLE
of the vaginal passage from the vulva. The flatter this is, or more nearly horizontal, the more difficult will be its negotiation by the dog, causing a lot of unnecessary fumbling and waste of energy, and frequently calling for assistance from the manager. The steeper the angle, the easier will be penetration by the dog. Exploration of this angle

will show how best the dog may be assisted, when circumstances appear to make assistance necessary, by either raising or lowering the bitch, or by giving the dog something for his hind legs to stand on. This will be more fully explained in the chapter dealing with methods.

2. THE LENGTH OF THE PASSAGE.

This varies. The extent of the vagina is determined by how soon the bony brim of the pelvis is encountered. If a short passage, the pelvic bone is felt with ease; if long, the extent of a little finger (in small breeds) or the middle finger (in larger breeds) may not reach it. The longer the passage the easier and more certain will be the service and the effecting of a "tie". If very short there is a tendency for a dog, after making penetration, to keep "slipping out" again, after only a few seconds. This is irritating, to the dog and to the manager, and is countered (immediately after penetration) by placing the hand on the dog's rump, firmly and forcefully pressing him to the bitch and maintaining contact for him. It may be necessary to keep the dog in this position for several minutes. Further, when the passage is short it is quite impossible for the large bulbular enlargement (mentioned in the anatomical description) to enter the passage and continue to enlarge therein, thus producing the "tie". Its appearance, however, outside the vulva does cause consternation to the complete novice, who probably did not know of its existence, or, if he did, where it should actually be at this stage. Provided that the forward portion of the erectile part of the penis is within the passage, and kept there by firmly holding the dog clasped to the bitch, he can be assured there is no cause for anxiety, that everything is going on satisfactorily, and that a service is being effected. Here an important question will be answered, since it is one so frequently asked. Is it necessary for a "tie" to be effected, and can a service be

THE SERVICE is meant to be:

considered satisfactory and conception counted on if no "tie" has taken place? The answer is that it is *not* necessary for an actual "tie" to be effected for conception to follow, and a litter of puppies to result. Provided there has been penetration, and the penis actually remained in the vaginal passage while ejecting was taking place, if only for a few moments and under one minute, this could be regarded as a satisfactory service, and the owner of the bitch be honestly assured that, provided both dog and bitch are fertile, a litter can be expected. The author learnt this lesson (since verified on many occasions) when himself a complete novice and using a stud dog for its first service. The dog was very raw, unintelligent and un-co-operative, and half an hour had elapsed before penetration took place, the dog remaining in position for less than a minute and then slipping away. Exactly the same thing happened two or three times more. The author was nonplussed, thinking that as there had been no "tie" it could not be regarded as a service and no fee should be taken. Fortunately, however, the bitch's owner, who was considerably experienced, was present throughout and had tried to assist, but did not appear to be in the least disappointed. He said he was quite satisfied that each time the dog was "there" for periods considerably under a minute, and one only a few seconds, a service had been given, and took the bitch away quite contentedly. The sequel—a litter of six puppies. As an example of one of the inexplicable vagaries of the service, the author has had the converse experience of a perfect mating and a firmly locked "tie" lasting one hour five minutes, during which all parties, human and canine, got thoroughly bored. Yet the result was a "miss"—no conception, no litter. For corroboration of these two extremes the author is indebted to an authoritative bulldog breeder, with two similar experiences. The first, a "tie" lasting three-quarters of an hour but with no puppies resulting. The second, an amazingly short period of actual penetration lasting only two seconds, the result being a record litter of twelve puppies! Having made a summary from a wide range of very diverse breeds, it can definitely be stated that the average duration of a normal service with a perfect "tie" is twenty minutes.

To revert to consideration of the dog immediately "slipping out". By such repeated abortive efforts to penetrate, the dog will sometimes swell completely and begin to eject outside the bitch. This may cause a novice to think that the effort has been wasted, that the service cannot be accomplished, and consequently to give it up. This is not so; it can be countered by removing the bitch altogether, and if the dog is a short-legged one holding him away from the ground to avoid contact irritation. After a few minutes the organ will retract into the prepuce. This can be accelerated by fanning with a newspaper, or applying cold water. The dog should then be returned to its kennel for half an hour. A further attempt can then be made, but it is in such a case as this that the "manipulatively assisted" or forcible method of mating has to be resorted to. The method is described in the relevant later chapter.

That such intimate details have had to be referred to is a matter of regret, but it must be remembered that the preface stated that no detail, the inclusion of which would assist a novice in case of unexpected difficulty, would be shirked on account of crudity. Any fool can drive a car at seventy m.p.h. on a wide, open country road, but it takes a wise man to drive carefully and slowly in difficult traffic. And anyone can conduct the majority of simple, straightforward services, but it is the wise man only who knows how to cope with the snags which crop up at odd times for any owner. The owner can become wise only from knowledge gained from an assortment of experiences from older hands. This book deals with the "snags", not with the normal procedure, which can be ascertained from abundant literature.

3. DIAMETER OF THE PASSAGE.

Sometimes this is so abnormally narrow that it almost precludes the admission of the little finger, and usually means a difficult union. In very severe cases dilation must be done previously by a veterinary surgeon. In many cases, however, repeated dilation with the little finger will make entrance possible, while at the same time the stimulation will induce a very shy, timid bitch to relax any muscular tenseness.

4. A STRICTURE, ADHESION, OR GROWTH

present in any part of the passage. This is felt as a hard obstruction to the exploring finger, beyond which it cannot go. In such a case a service is usually quite impossible. The owner of the bitch should be told and no fee should be taken as no service has taken place and there is therefore no possibility of conception. He should advise the owner to take the bitch to his veterinary surgeon for advice. This precludes any false statement a bitch owner could make to the effect that the stud dog could not mate the bitch or was not potent. In the few instances when we have found it quite impossible to effect a service, and the bitch owners have been so informed, we have been fortunate in that the veterinary surgeon has confirmed that an anatomical defect existed.

5. SOMETIMES THE BITCH IS BROUGHT TOO LATE;

the vulva has tightened and is very small and the lumen (width or diameter of the passage) has diminished. Quite frequently the vulva of some bitches remains tight and small throughout the whole oestrum, although they will evince "readiness". This must not be confused with (3) above, an anatomical narrowness while the bitch is still "very ready" in other respects, though the conditions are almost identical. By the application of vaseline, by dilation and stimulation, a service can sometimes be managed with a tight vulva, but frequently it cannot.

The owner must then be informed that no service has taken place, and the reason given. No fee is justified.

Although it has been said that many breeders make a regular routine of digital exploration for the purposes described above, many others consider it necessary only when they come up against some difficulty that makes penetration difficult or impossible.

The principles having been explained, it can now be left to the reader to discriminate, and work out his own procedure after experience. The author would sum up by saying that digital examination is by no means always necessary but is frequently highly desirable. Whenever a digital examination is made the following preliminary precautions should be taken: the nail of the finger should be cut very short, and be smooth. The finger should be thoroughly scrubbed and immersed in an antiseptic solution, e.g. dilute Dettol or T.C.P. Before insertion it should be well smeared with medicated vaseline.

THE DOG.

Compared with the bitch, in the dog the presence of any anatomical abnormality of the sex organ is extremely rare; for all general purposes it may be regarded as normal. There is, however, a fact connected with the practical application which should be noted; it is the flexibility of the organ which enables the dog to be turned completely round to face the opposite direction when "tied". Among breeders this is known as "turning" the dog. The method of carrying this out is described in a later chapter, but the following is an explanation of the reason for the procedure. It is always observed in practically all breeds, particularly those of the more aggressive types as compared with the more domesticated, that when a dog has mounted and is serving a bitch he becomes restless, fidgety, not at ease in the position, and exhibits a desire to "dismount". If he does so before the bulb swells in the passage, this is a cause of the tiresome "slipping out". If actually tied and a dog tries to dismount, it is one of the situations which made necessary the information that a dog can injure himself and the bitch during the service, and at which stage attention is necessary to prevent such injury. In a perfect service the lock is so firm that it is almost impossible for the dog to disengage himself; but a fractious dog endeavouring to get free will tug and struggle, to the discomfort of himself and the bitch. He must be gently but firmly restrained until the swelling has reached the maximum and the tie completed. He can then be "turned", and it will be found that the "restlessness" ceases and both dog and bitch quieten down, each appearing to prefer this "turned" position and remaining steady till the service is completed.

There is a theory—and it is only a theory—to explain why a dog prefers to be turned. The dog at some time lived in the wild state. While carrying out a service, with his front legs clasping the bitch, both are completely defenceless if attacked. The instinct doubtless remains; if other dogs approached while a service was in progress they would be regarded as rivals by the dog, and at a time when he would be vulnerable to attack, as also would be the bitch. If, however, he is dismounted during the tie, and each is facing in opposite directions, each would be able to put up a fight. To what extent this theory is true is difficult to judge, but it is certainly a routine with breeders to allow the dog finally to assume the position by "turning him", having found that he will then stand passively for the duration of the service. When completed, separation is simple and natural; some breeders think it advisable to hold up the bitch's hind legs for a short time to prevent the semen running out, but it is unnecessary to do so. The bitch should now be returned to her kennel and the dog led off.

CHAPTER II

Responsibilities of the owner of the bitch and the stud dog owner;
the obligations of each to the other—Accommodation for bitch—
Preparation of a mating-room and its equipment—Arrival and
return of bitch—Fixing the date—Mode of despatch of bitch—
The travelling box—Labels—Sanitary and hygienic condition of
the bitch—Number of matings.

AT THIS stage a digression from anatomy and intimate
detail may be refreshing. Consequently, before dealing
further with various methods of mating we will consider
certain preliminaries which are the responsibility and
the mutual obligations of the owner of the stud dog and
the owner of the brood bitch.

THE STUD DOG OWNER.

1. *Accommodation for bitch*. If the establishment is a
modest one of one or two dogs only, but advertises a dog
at stud, or is one of the most extensive Kennels, suitable
accommodation should be reserved for housing a bitch
who may be a visitor for a day or two. It is advisable first
to ascertain if the visiting bitch is accustomed to being
indoors as a house dog or to living outside. Some owners
take the trouble to allow a "house bitch" to be accom-
modated indoors in a room set apart, the door of which
can be safely secured. Otherwise a separate kennel outside should be used, and permanently kept for visiting
in-season bitches. It should be in some secluded spot, as
distant from other dogs or bitches as possible. A small
run should be attached, preferably covered in, every care
being taken that there is no possible chance of the bitch
getting out, or of a dog getting in. It is a prime responsibility of the stud dog owner that no *mésalliance* shall occur
while the bitch is in his charge.

Fresh clean bedding, straw or wood-wool should be placed in the kennel for each new arrival. The old bedding should be burnt as soon as a bitch departs, and the kennel should be scrubbed with disinfectant. In case the bitch has left any "undesirables" behind it is advisable to sprinkle the kennel with an insecticide powder such as D.D.T.

2. *Preparation of a mating-room.* The size of the room in which matings will be carried out is immaterial; generally, however, the smaller the better. An outhouse, small shed or empty garage are suitable; or the room can be used in combination with the one where grooming is usually done. Essentials are that it should be as bare as possible, and it should have good light, both artificial and from a window. The latter is useful to look through from the outside when a dog and bitch have been left alone together for a few moments in the early stages. An electric heater or some other form of heating, suitably guarded, is useful in the winter months. A secure latch, or lock and key, is essential.

3. *Equipment in the mating-room.* To cover all the methods of mating described in Chapter IV the following equipment is necessary. A table approximately 2 ft. by 4 ft. Attached to this at one end it is useful to have a post into which a large staple has been driven, or an iron ring screwed in. Alternatively, if one end of the table is pushed against a wall the staple or iron ring can be fixed to the wall. This meets the requirements of the "hitching-post" method. One or two low stools or hassocks, or a low chair, or a bale of straw or a rolled-up rug; these are for the convenience of the assistants in the "on-the-ground" method. A strip of coconut mat on the floor, or other "non-slip" covering, for the convenience of the dog, and also a rolled-up sack to assist him when it is found necessary to raise his hind legs to a bitch much taller than he is. Other items of equipment include: vaseline for smearing round the vulva to facilitate

penetration. A strip of 1½-in.-wide bandage from 2 to 3 ft. long according to size of breed. When it does become necessary to muzzle the bitch this is by far the quickest and simplest method; its use is described later. A pair of thick leather gloves, as a precaution with an obstreperous bitch who attempts to bite. Similarly, in the exceptional case of a human being bitten some tincture of iodine, or T.C.P., is useful to have handy as an antiseptic before applying a dressing. A pair of scissors to cut long hair from the hind legs and in the region of the vulva in the case of long-coated breeds. (See further note later.)

4. *On arrival of the bitch.* As the bitch will have had a long journey delay should be minimized by either being at the station to meet her or arranging that her arrival is at once notified by telephone. Take her box straight to her kennel or isolated enclosure, use her pet name, and let her out of her box after making certain that all gates are secured and that her kennel or enclosure latch is safe. If her own isolated run is small, put a lead on, carry her some distance, then let her have a lead walk or run in a suitable spot. On return offer her a meal and a drink; if exceptionally nervous and highly strung, include a quarter to half a grain phenobarbitone, according to breed. Examine the vulva and decide if she must be mated the same day. If so, delay doing so till she has been bedded down for a few hours. If it is quite definite that she will hold till the next day settle her down for the night. See that any detachable gadgets for securing her travelling box are put in a safe place and not lost before she is due to return, and fasten her own lead to her travelling-box. Empty and clean her travelling box and replace fresh wood-wool or straw. By 'phone or wire inform her owner of her safe arrival.

5. *Returning the bitch.* If only a short journey and her arrival was early in the morning she can sometimes be mated and returned the same day. Almost invariably,

however, she is kept at least one night, possibly two or three. By 'phone or wire inform the owner of the bitch the result of the service, and give full information regarding the return journey. Dogs are frequently mislaid while travelling owing to carelessness en route. Consequently, the following details should be given: Time and station of departure; stations at which changes will be made and times of departure therefrom; time and station of arrival. This information helps the owner to trace a dog which does not reach its destination when expected. Exercise her before placing her in her box and offer her food. Ensure that her box is securely fastened. Write clearly on the label the name and address of owner, and it is advisable to add the owner's telephone number. If the bitch is to be called for at the station, and not delivered, state this boldly in block letters. One label at least should be fixed by drawing-pins—tie-on ones get torn off and sticky ones frequently don't stick. If you have no printed tally bearing the words, add another large label, VALUABLE DOG. DO NOT DELAY. A word of advice about bitches flown from or back to Eire. Firstly, they travel extremely well by air and are in no way upset by the experience. It is infinitely the best and quickest way; if you are not too distant from an airport the whole journey can be made in a minimum of two hours or maximum of five, while frequently cross-country journeys in England, involving changes and missed connections, can take twelve or fifteen hours. Secondly, airport officials seem to make a practice of removing a bitch from the basket; it is well-intentioned but it is disconcerting to see your in-season bitch being brought to you gaily trotting along on a lead—your heart misses a beat if you think that the collar might be loose; a bitch in strange surroundings might tug and slip it. On at least two occasions a kindly air porter has brought a bitch to us walking on its lead, and in Dublin it seems the common practice. To obviate this we now add yet another label for bitches travelling by air: PLEASE DO NOT REMOVE DOG FROM TRAVELLING-BOX, EXCEPT IN EXTREME EMERGENCY.

In air travel no wood-wool or straw is allowed in the box. Sometimes rugs have to be removed, only paper being permitted.

One final word on travel generally. If the journey is likely to be a long one with changes it is better for the bitch to travel overnight. The dog, having been active all day, invariably sleeps through the night journey and arrives much fresher, and conditions are quieter at stations where changes are made and there is less risk of missed connections.

RESPONSIBILITIES OF THE OWNER OF THE BITCH.

1. *Fixing the date, and mode of despatch for the mating.* Having decided on the stud dog to be used, the owner of the stud dog should be given early notification of the proposed date of mating; hence the importance of the first sign of "showing colour", the date of the service being fixed from this day. Usually this is eleven days afterwards, unless the best mating time for a particular bitch is known. If the stud dog owner is careful not to over-use his dog he may refuse to accept a bitch at short notice; his dog may have served at too short an interval beforehand. Therefore the service should be booked well in advance. The next point to decide is whether to take the bitch to the dog or despatch her by rail. Distance and circumstances govern this, but whenever the bitch can be taken by the owner it is more satisfactory to do so. The bitch is less nervous, the owner can see the actual mating and assist if the stud dog owner wishes it. Further, the business side of the transaction can be completed on the spot. As it is impossible to predict how long the mating will take it is inadvisable for the owner to have other engagements that day. When it is impossible to accompany the bitch the stud dog owner should be given definite instructions and information regarding the train journey,

notifying him of the time of departure and time of arrival in order that the bitch can be met without any delay. A 'phone message or wire on the actual day, confirming despatch, is useful.

2. *The travelling box.* A strong travelling box should be used. British Railways' rules are that it must be big enough for the dog to stand up and turn round in, this applying to any size breed. It should have good ventilation holes, but it is preferable for the box to be enclosed so that the bitch cannot see out, this being less disturbing to her when waiting on a crowded platform. It must be well secured, but whatever locking device is used it must be such that it can be opened in an emergency during transit and by the stud dog owner on arrival. Methods involving nuts and bolts which can be lost are undesirable, as also are split pins. Quite the simplest method, giving completely adequate security, is the following: the door or lid has an ordinary gate latch or loop which fits over a staple or ring which is *fixed* to the box. The "scissors-grip" catch end of a dog lead, with about three inches of the leather, is fixed to the box at a distance that will allow the grip to engage the ring when the loop is over it. Thus *all* parts are fixed to the box and none can be lost, and attaching the lead catch to the ring staple is a matter of a second only. The bitch's own rug or blanket (of which she knows the smell) should be placed inside; alternatively, some clean straw or wood-wool. A collar and lead should accompany the box but not be worn. A bitch rarely wants to eat when travelling, but a handful of small biscuits should be included in case she does on a long journey.

3. *Labels.* Advice on labels is the same as that given in previous section (5), dealing with the return of the bitch by the stud dog owner.

4. *Condition of the bitch.* It should be a matter of personal pride on the part of the owner that the bitch is sent in the best possible condition. For her nuptials she should be brushed, groomed, bathed, shampooed, or

whatever is the customary method for the bitch to be turned out in an immaculate condition. It is the height of bad form, and most inconsiderate to the owner of the dog, to send a bitch who is dirty, her hind parts unwashed, and her hair there matted. In the case of long-haired, heavy-coated breeds, it is usual to cut away all the hair in the vicinity of the vulva, the rump and upper parts of the hind legs. This not only "tidies up" the bitch but assists negotiation by the dog. A mass of hair would act as a buffer and impede penetration. It is usual for this to be done before the bitch is despatched, since the stud dog owner should not do it without permission. If it becomes essential for him to do so he should use his discretion but inform the bitch owner of the necessity for his action. In any case the bitch will be off the show bench for four or five months and the lack of hair will not be detrimental during that time, unless she has show engagements to fulfil during the first four to five weeks of her pregnancy. Thereafter she is usually unshowable and loses her coat at whelping. The cutting away of the hair is also beneficial at actual whelping time. To despatch a verminous bitch is a disgrace; to send one infected with, or one who has been in contact with a case of, distemper or hard-pad is a crime. Perhaps this is the worst offence that can be committed in dogdom; it is certainly the most important responsibility and obligation of the owner of the bitch to the stud dog owner. "Do as you would be done by." Imagine your own chagrin if you had your valuable Kennel of show and breeding stock wiped out as a result of accepting an infected or "contact" bitch. Stud dog owners are quite within their rights to refuse to mate a bitch in unclean or verminous condition, and are justified in immediately returning it to the box and putting it on the next train back. Such cases have occurred and will occur. As a bitch owner, guard your reputation; if the bitch is worth mating she is worth grooming. However keen you are for a litter, and whatever the loss would be, sacrifice your own ends rather than risk loss to another breeder by sending him an infected bitch.

5. *Number of matings.* Some bitch owners request or insist that two services be given within an interval of two days; in such cases the owner of the dog must comply. If, however, a first service was satisfactory in every way, and both animals are potent and fertile, a second service is entirely unnecessary. It is no more likely to induce conception than the first one was, and it has a definite disadvantage in that the complication of a double conception may arise—that is, actual conception and fertilization at each of the services. If there is the slightest doubt about the adequacy of the service an interval of a day or two can elapse and a second service then be given, when the bitch may be more "ready" and other conditions causing the doubt be more propitious. Again, if a bitch is undergoing a course of hormone treatment for fertility, or increased prospects of conception, under a veterinary surgeon he usually specifies days and times when matings shall be done, and sometimes requires the bitch to be mated three times—once very early, a second at the average time, a third very late. The bitch owner should give details of the services required, with particular instructions in the case of a bitch under treatment. With these the stud dog owner should comply.

6. *Return.* It is courteous for the bitch owner to notify the stud dog owner, by 'phone or wire, of the safe return of a bitch.

CHAPTER III

Idiosyncrasies in the behaviour of a dog or bitch at mating time—Stud dog behaviour—Reactions of the bitch—When necessary to muzzle the bitch; procedure.

THE DOG.

THE behaviour, mannerisms and peculiarities of dogs at mating time are very variable, therefore the novice must accept at the outset that hard-and-fast rules for management cannot be laid down. He must appreciate that he must learn by observation the individual mannerisms of his own dog or dogs, and into which class each can be placed. One rule that can be laid down, however, is that whatever idiosyncrasy a dog shows at its *first* service, so will it *always* behave, and must always be allowed to behave thus and be treated accordingly. It can be regarded as belonging to a certain type. From general experience it is considered that a dog seldom changes its original technique. Once a dog's method has been determined it is easy to know how to manage him.

TYPES.

To classify stud dogs into types one would say there are four:

1. *The aggressive* "CAVE-MAN", "HE-MAN" *type*. This dog knows exactly what the bitch has come for and wastes no time on preliminaries. He will "mount" immediately, and if the bitch is at the optimum period of "readiness" and is co-operative he will negotiate easily. Penetration is effected in a flash—like a "trout up stream". Provided the dog is not too aggressive, lucky is his owner as many difficulties are removed. With this type, however, one must be careful with a highly strung, nervous, non-co-operative bitch. The dog must be

restrained a little; holding him on a lead, induce him to be more gentle and at least give the bitch time to appreciate the situation and become amenable. If this type is allowed immediately to serve a nervous bitch she tends to become still more frightened. Her muscles tighten instead of relaxing, making penetration by the dog much more difficult. She will snap, or actually bite, doing all she can to avoid the union by turning and pulling herself away or rolling on the ground. Gently restraining the dog, and patiently soothing the bitch, will in the end bring about the desired result, but assistance must not be hurried; great patience is required.

2. *The* "AVERAGE" *type* of dog is neither too premature and aggressive nor too sluggish and unintelligent. Instincts may develop slowly; he will want to take his time, and will sometimes move away, sit down and "think it all over", then come back to the bitch with renewed eagerness. He usually wants to demonstrate his affection by nuzzling his head against her and licking her ears. He is in no way put off by the coquettishness of a "ready" bitch, who will run around, then stand, and then run about again; he plays up to it and prefers this preliminary flirtation to the downrightness of the "cave-man". This type is eminently suited to the natural or "love mating" and, provided the assistant does not try to hurry either dog or bitch, this can be regarded as the normal, straightforward, simple service. In ordinary circumstances, we strongly advise the "love mating", and the more natural the service can be kept the better. It is only when idiosyncrasies occur, and difficulties arise with either dog or bitch, that its practice should be departed from. But it calls for time and patience on the owner's part. He must give the lovers opportunity to decide for themselves when they wish to make the union complete, and only then intervene in certain ways to assist the mating. The methods of assistance are discussed in the next chapter. Again, the owner who possesses this "average" type dog is

fortunate; although a bit slow, his dog makes a sure and sound stud dog and gives very little real difficulty.

3. *The "JITTERY" type.* Some dogs are nervous, shy, difficult, timid and fussy; they appear to be overawed by the proceedings. If the bitch becomes skittish and chases such a one he will run away, not in the playful mood she is exhibiting but from stage fright at her unseemly advances. When a degree of excitement has been created he will diffidently mount, but drops away again. He peers apprehensively at the bitch, and is put off by the slightest discouragement. He is the complete opposite of the "cave-man". Even when mounted and doing his utmost to serve her, if the bitch is snappy and is allowed to turn her head, show her teeth or growl (as some bitches do) his ardour is completely quenched; he drops away and sometimes won't "play" again. When this type, after exhibiting all the jitters mentioned above, does eventually penetrate he must at once be pushed firmly up to the bitch and kept there, otherwise his fussiness will cause him to slip out again as soon as he has negotiated. Further, even when tied and functioning he continues to jitter, is restless, and when "turned" will "pither" about on all four feet. He tries to come apart or wriggle, which causes an occasional cry of pain from the bitch. With this type there is unquestionably a need for two assistants. A single-handed mating here can be very distracting and irritating, but with two persons it can be entirely simplified. One would regard this as the most difficult type of stud dog, but if he is by conformation and points a grand specimen and throws good progeny it is obvious that he must be persevered with, humoured, and gradually trained.

They may improve a little but they never completely shed their diffidence. They will never become "he-men". If your stud dog is of this type you must fervently hope that all bitches coming to him will be docile, ready and co-operative. If they are not, the difficulties are accentuated. But "difficulties are made to be overcome" and these can be countered. Have two persons ... coquettish the bitch may be, this typ... responds to the "natural mating". Theret... time by expecting him to react normally an... the bitch an opportunity to "flaunt her b... before him. She should at once be held station... straight to the front, by one assistant placing a ... each side of her head and neck, grasping her rea... ably firmly so that she cannot turn round or even move from this position. That is one assistant's job throughout the whole proceedings—to keep the bitch still and prevent her turning her head. If she is ready and co-operative and "curling her tail" the dog will in due course attempt a tentative mounting; let him do this a few times as stimulation usually increases with each. The other assistant should be on the alert to watch for the first actual negotiation, and then immediately press the dog closely to the bitch and very firmly hold him there.

With this type it is very inadvisable to attempt to "turn" the dog too soon; often it is better not to do so at all. Hold him in close contact with the bitch throughout and let him jitter and dither there; it makes for a more certain conception with less risk of hurting the bitch. It is tiring for the assistant, but necessary. If the dog is too diffident even to mount and attempt penetration himself, he is the type which requires the "guided" or forceful method, described later. The attitude of this type is not due to lack of sexual instinct, which may be quite keen, but to lack of confidence and marked shyness.

4. *The* "SLUGGARD". In all breeds some young dogs are sluggish, lethargic and quite unintelligent as far as the service is concerned. They are sexual nit-wits. They are similar to the shy and nervous type, the only differ-ence being that in the former the sexual instinct is normal, while with the "sluggard" it is underdeveloped or lacking. This is the only type where changes in be-haviour do ultimately occur; as the dogs get older, and

th each service accomplished they show improvement. They are the most difficult type to deal with, but fortunately they are only a very small minority.

It is necessary for the assistant to resort to hand stimulation by rubbing the penis and, if response results to the erectile tissue, guiding the penis to the vulva and actually inserting it, which is the "forcible" mating. In time this practice educates the dog sexually and it will gradually become less necessary to employ it. One has experience of such a dog eventually becoming quite a sound and "average" stud dog, but his first half-dozen or so services called for a lot of patience. If the first is at about ten months, and such a dog appears to be quite impotent after trying all our resources, it is better to postpone further attempts for about three months, and hope that further development will reduce the sluggish tendency.

Hand friction is sometimes required to a slight extent with types 2 and 3 also, if they appear to be uninterested. In many cases it has the desired effect but some dogs resent it completely, and it has the opposite effect, putting the dog off entirely. This is an example of the necessity of knowing the dispositions of each of one's individual stud dogs. Another example: although it has been mentioned that sometimes it is necessary to assist a dog by pressing him to the bitch, some dogs will be put right off if touched at all. Two further general remarks concerning classes 2, 3 and 4 are, first, that the dog should be encouraged by using his name in a praising tone—he should know that he is acting with your approval. (This because they have sometimes been scolded for acting on a natural impulse on occasions when quite definitely they should not.) Never be harsh with a young, untrained stud dog during a service; to be impatient defeats your object. The second suggestion may sound rather too simple to mention. If a dog is being very dilatory, and not "getting on with the job", pretend to take the bitch away by lifting her up, if a small breed, or leading her out of the room if a large breed. At the same time use such phrases as, "We'll take her away," or, "We'll send her back." When there are several stud dogs available (they invariably know what is afoot and usually some jealousy is apparent), and the selected one appears uninterested in the bitch, say, "Oh, we'll use Bob instead and take Bill back," and pretend to lead Bill off. The use of such threats and actions always increases excitement to a surprising extent, and it will be found when the bitch is put on the floor, or brought back again, the dog will mount at once and negotiate quickly. Such simple advice may appear almost childish, but anyone with experience amongst dogs will agree that they do uncannily understand our "language tone" and our actions. As this procedure has so frequently served its purpose and simplified a service it is worth trying.

THE BITCH.

Compared with the dog, "mannerisms" in the bitch are considerably fewer, making any classification into types unnecessary. Provided one has been able to time the service with the period when the bitch is at her most "ready" stage and "curling her tail", practically all will stand quite willingly and passively to be mated. The majority of bitches are of this type. It is possible that at the time of actual union bitches have a much keener sexual instinct than young dogs "under instruction". Bitches very seldom exhibit their sexual instinct at any time other than at the oestrum, while dogs evince it frequently at other times but appear to be less keen than the bitch at the actual time.

Some maiden bitches are very scared, partly due to strangers and strange surroundings, partly due to fear and the roughness of a dog. A few are inherently snappy and even vicious. A bitch is always more docile and less scared or anxious at her second and subsequent matings. On introducing a maiden bitch to the stud dog you should quickly decide how she is going to react. If she is coquettish and plays at once you can expect her to behave like the majority of bitches, and the service will be simple and straightforward. If she instantly runs away, trying to hide behind something, or stands at bay, snarls or shows her teeth to the dog, this is due to fear, which is not overcome by leaving the dog and the bitch together in the hope that they will mate naturally. Further, if the stud dog is young and inexperienced he can be completely put off stud work for some time if the bitch continually snaps and growls at him, and an otherwise promising stud dog's future is marred.

When bitches are perhaps a day previous to their best "ready" stage, or perhaps a day past it, they are still perfectly mateable. Generally they are more willing and co-operative after than before. In such circumstances the bitch will frequently not be responsive and co-operative but will snap and show resentment to the dog. In this case, as with the frightened maiden bitch, the manner of handling them is the same. They should be addressed in a soothing tone, very gently and coaxingly, using their own names but at the same time be firmly made to stand, and be held in that position.

It is with the snappy or frightened type of bitch that muzzling is strongly advocated. The following method is recommended as it is simple, rapid and perfectly effective.

Having handy (as already mentioned under Equipment) a length of $1\frac{1}{2}$-in.-wide bandage, make a single loop in the middle with the ends downwards. Slip this over the muzzle; pull tightly but not enough to cause pain; bring the ends up and cross them over the muzzle and then downwards; repeat this, pulling reasonably tightly each time; then tie under the chin, then cross the ends under the neck and tie in a loose bow over the neck. It is amazing how rapidly this can be done, and how it seems at once to quieten a fractious bitch. Securing the ends over her neck restrains her from turning her head and considerably restricts any movement of the head at all. In necessary cases, probably quite infrequent, it is a complete protection for owner and dog from being bitten by a frightened or vicious bitch, and at the same time the bitch becomes more tractable. Other antics of a "not quite ready" bitch are attempting to sit or lie down, or a determined effort to roll on her back.

It is with this occasional difficult bitch behaviour that it is most advisable for there to be two people to assist at the mating.

Both dogs and bitches can be quickly put off, sometimes at a crucial moment, by an unexpected noise, a sharp voice, too much loud conversation between assistants, particularly if one or both are strangers. Consequently, when possible a quiet and secluded room or site should be chosen for the mating, where the barking of other dogs will not be heard. Conversation should be reduced to a minimum and be carried on quietly. The door should be locked to prevent intrusions; this is mentioned because one has had experience of dogs dismounting at preliminary stages because the door has been unexpectedly and noisily opened.

An attempt has been made in this chapter to cover all contingencies a novice may meet in the behaviour of dogs or bitches at mating time. Stress is again laid on the fact that most services are simple, straightforward and trouble-free and form an interesting part of the dog-breeding hobby. The information should not confuse or deter the novice but, rather, place him in a position to cope with a difficulty as it arises. He should not now be caught unawares by unexpected behaviour of either dog or bitch. To be forewarned is to be forearmed.

CHAPTER IV

Methods of effecting a mating—The natural mating—The assisted mating; single-handed, and with two assistants—The table method—The hitching-post method—The "guided" or forcible mating.

IN THE last chapter the reactions of dog and bitch at mating time were considered. In this chapter will be described various methods employed to effect the service. The variations in methods are due to the idiosyncrasies already mentioned; they are also necessitated by the varying sizes in a range of over ninety breeds. The author has not confined himself to breeds of which he has had practical experience but has sought information from experts in breeds which by size and type could be grouped. It is hoped that guidance will in consequence be found available for the novice whatever the breed of his choice. Each experienced breeder has at some time evolved methods which suit his own ideas, his circumstances and environment, the characteristics and temperaments of his breed or breeds, and by which he has found he gets the most satisfactory results. In the same way the novice will gradually accustom himself to the method which best suits his requirements. Obviously no single method can be advised as the only right way to manage a service.

THE NATURAL MATING.

When a stud dog is intelligent, and of the normal average type, and the bitch is "ready", willing and responsive, they should be left together for Nature to take her course in whatever manner the pair prefer. By this method there is minimum interference, the assistant standing on one side and taking no part until penetration is on the point of being, or has been, made. Then it is

68

advisable to steady the bitch by holding her head and shoulders facing square to the front. Up to this stage, some breeders go out of the room and watch proceedings through a window, finding that the animals will more readily react if left alone. They return to the room when negotiation is imminent. The only other assistance required is given when one is quite certain the tie has been made (ascertained by feeling under the bitch and noting that the swelling has taken place inside); then the dog can be turned. No attempt should ever be made to turn the dog if the swelling has taken place outside. This can be regarded as the simplest and most normal service, no other interference being necessary than that described. It is the opinion of all breeders that in no circumstances should a dog and bitch be left completely unattended, as has been done by amateurs feeling that such a natural event could be left entirely to the participants. There is a risk of injury to both dog and bitch if this course is followed.

Reference to "turning the dog" has now been made twice, but the actual operation has not yet been described. When the tie is definite, and there is no risk of separation, the dog's front feet (usually clasped round the bitch if he has not dismounted) are gently placed on the ground on the right side of the bitch, the two then facing the same direction and standing side by side. The dog's left hind leg is then gently raised and carried over the bitch's back to the right side while the dog himself turns in the direction in which the bitch's rear is facing. The left foot is then placed on the ground, while the right one has moved towards the left side. Thus the left foot now stands where the right did, and the right foot where the left did, with the feet pointing in the completely opposite direction. It is a sound plan to grasp both animals' tails and hold them firmly together in one hand, if they are of sufficient length; this prevents any attempt at dragging away until the right time. It is astonishing how both dog and bitch seem to prefer this turned position, remaining steady and passive.

THE ASSISTED MATING, SINGLE-HANDED.

We will now consider a situation in which a stud manager receives a bitch for mating, has no one available to assist and must supervise the service on his own. There are two possible methods, detailed in order of preference. They are most suitable for the smaller breeds up to Fox Terrier size and particularly so for the short-legged breeds, such as Corgis, Dandie Dinmonts, Scotties, Cairns, West Highlands, Dachshunds, Sealyhams, etc.

(a) *On the Floor*. The manager can use a hassock, a low stool, a bale of straw, or a rolled mat, on which he himself sits, about six to twelve inches above the ground, with his legs wide apart and knees slightly raised. He places the bitch across his left leg below the knee, and nearer to or farther from the knee according to the height of the breed. The bitch's belly is thus supported by the leg; the height of the vulva can be raised to the dog by drawing up the knee, or can be lowered by dropping the leg and stretching the bitch's hind legs outwards. The dog is brought inside the right leg and either himself mounts or is placed in the mounting position. Mating is then allowed to take place. The advantages of this method are that the manager, while gaining a large measure of control over the bitch's position with his left leg, has both hands free. Thus, whether the bitch has been fractious and was muzzled, or had only a collar and lead on, the manager can with his left hand easily hold the lead or muzzle, or support the neck and shoulders. The main object of this procedure is to keep the bitch's head facing front and prevent her turning round. At the same time the right hand is free to give the dog any assistance that may be required, and to keep him pressed close to the bitch to prevent "slipping out" in the early stages.

In the case of the large breeds and the long-legged dogs, such as Setters, Mastiffs, Danes, Borzois or Irish Wolfhounds, it is obvious that this method would not be practicable. I have learned that it is extremely rare among the big breeds for anyone to attempt to control a service single-handed, that as the breeds get bigger the breeders always try to have a second person present. When a single-handed service with a big breed is unavoidable the manager should stand up, and raise one knee with his foot resting on a stool, thus supporting the bitch's belly over the knee.

(b) *The Table Method.* There are many breeders among the smaller and low-to-ground breeds who prefer to work on a table rather than on the floor. In such cases it is still advocated that with a co-operative pair the preliminaries of the "natural mating" should be allowed on the floor, and the animals not placed on a table until stimulation has occurred for both, as a result of some acquaintance. It is obvious that any form of table method is quite unsuitable for most of the big breeds.

THE HITCHING-POST METHOD.

As an adjunct to the table method there is the "hitching-post" method. This consists of securing to one end of the table a piece of wood, slightly higher than the shoulder height of the dogs, and fixing into this a strong staple at shoulder height. Alternatively, the table can be placed firmly against a wall and the staple fixed at the right height into the wall. By tying the bitch's lead close to the staple, allowing only about six inches of lead from collar to staple, another effective means of controlling the bitch's head and preventing her turning round is gained. This method leaves the hands free when one has to supervise a mating by oneself, when both hands are required to manipulate the vulva and the penis, and adjust the one to the other. Generally the bitch stands quite passively when controlled on a tight lead in this way, and it is

impossible for her to snap at, bite or in any way put off a
raw young stud dog under "service training".

In single-handed matings among the big breeds a
staple in a wall at the shoulder height of the dog when
standing on the floor is the alternative to the table
hitching-post method for the small breeds. The bitch is
secured by it close up to the collar, thus giving the manager
more freedom of manœuvre at her rear end when he need
pay no further attention to her head end. With the big
and long breeds, such as the Deerhound, Wolfhound or
Borzois, this advice will be found most helpful.

Choice between the floor method and the table method
is often determined by the stud dog himself; some will
mate readily on the floor but refuse to do so if placed on
a table.

The way in which a dog first serves or is trained to
serve will be the way he will always prefer.

THE ASSISTED MATING WITH TWO ASSISTANTS.

(a) *Small and Low-to-ground Breeds*. When two people
are able to take part in the management of a service it is
considerably simplified and can be carried out with
greater certainty, efficiency and comfort for the dogs.
Mention has already been made of the difficulties which
may be met with, and of certain actions by the dog and
the bitch which must be prevented. These can be re-
capitulated thus: (1) A bitch being scared and running
away from the dog; (2) A dog being nervous and continually slipping out after penetration; (3) A bitch sitting
on the floor or rolling on her back at each attempt
of the dog to mount; (4) A bitch snapping at and biting
the dog on first introduction; (5) A bitch repeatedly
turning her head and straining away when mating is first
effected. Any or all of these things can tend to put off a
young stud dog and perhaps permanently ruin him for
the work. They can all best be avoided by having two
persons to assist at the service. Each should have a definite

job to do and keep to his own job. Responsibility can be quite simply divided: one attending to the head end of the bitch, the other to the rear end and to the requirements of the dog. Each is of equal importance, and together they can prevent the untoward happenings already mentioned.

With the small and low-to-ground breeds the following is a routine procedure: we will call the assistant at the head end "A" and the one at the rear end with the dog "B". Assuming that the bitch is one of the fractious type, and that muzzling is decided upon, "A" proceeds to fix the bandage muzzle, while "B" smears the vulva with vaseline, and makes a digital exploration for the reasons already given. These preliminaries ended, "A" can either kneel on a cushion or sit on a low stool, and grasp the bitch's head, neck or shoulders on each side firmly, at the same time speaking to her soothingly. His main purpose is to prevent her from suddenly swinging round. This is most likely to occur at the actual moment of penetration, when the bitch frequently gives a sharp cry or whimper, and it is for this that "A" must keep an alert watch to anticipate it, and exert firm control just at this stage. Depending entirely on what may be expected in the temperament of a bitch, "A's" job may be simply the quiet one of a lightly restraining hand on the bitch's neck whilst caressing her with the other, if she has a normally tranquil temperament. Or the job may be a regular wrestling match if she is of the supersensitive type, highly strung, scared and vicious, and in such cases he does his job correctly only if he has held the bitch so firmly that she has been unable to swerve round suddenly, causing the dog to slip away, or in any other way interfere with him or put him off the performance of his natural function.

According to the size of the breed, "B" can either assume the same sitting position as in the single-handed mating, having full use of both hands, or he can kneel on a cushion or sit on a low chair. "B's" main jobs are, firstly, to decide whether the bitch is standing too low or too high for the dog, and adjusting the bitch to him by raising or lowering the vulva with the finger of the left hand. If the bitch is too high her hind legs should be stretched well apart and the dog given a folded rug to raise his hind feet. If the bitch is too low, she should be given the folded rug for her hind feet to rest on. Secondly, having thus assisted negotiation "B" should be ready at the moment of penetration to press the dog's rump firmly to the bitch as this can be regarded as the most critical phase, because of the tendency of the bitch to swerve and of the dog to slip out. Instantly the dog will himself start working and if his idiosyncrasies will permit we prefer to maintain the pressure by "B" during the dog's exertions, relaxing it only when the dog's efforts are over and it can be determined that the swelling has taken place inside and a tie effected. Thirdly, "B" is in the best position to turn the dog when the right time comes.

(b) *The Large and Tall Breeds.* As one moves up the scale of breeds taller than Fox Terrier size, and longer than Dachshunds and Dandies, modifications of method will become necessary for the two assistants at the assisted mating of the bigger and longer breeds up to the Danes and Wolfhounds, and for these intermediate sizes the following are suggested methods:

(1) A stool or low chair is placed on the floor beside the bitch. "B" places his left foot on this, raising his bent knee until it supports the bitch's belly. "A" concentrates only on the bitch's head and shoulders, as with smaller breeds. Thereafter all the previously described general principles are observed. With an obstinate bitch disposed to sit, lie or roll, both "A" and "B", standing on opposite sides of the bitch, place a foot on the stool or chair alongside each other and with the sides of their knees in close contact can make a bridge under the bitch. Each has two hands free to attend to his department. This method of "encirclement" gives a good measure of control as, in addition to pressure of the assistants' bodies against the

sides of the bitch, she can be adroitly manœuvred from underneath. It is helpful with the heavy breeds, such as Bull-Mastiffs and Danes.

(2) A further modification for the biggest and heaviest breeds is for "A" and "B" to stand opposite each other with the bitch between them, "A" a little nearer the head end and "B" the rear end. "A" places his left hand under the bitch and "B's" left hand is likewise placed there, each grasping firmly the other's wrist. Here again a useful controlling and encircling bridge is made. Control here is even better than with the knees, but each has only one free hand, either for "A" to keep the bitch's head straight or for "B" to maintain pressure against the dog's rump.

THE GUIDED OR FORCIBLE MATING.

In addition to the methods and modifications already mentioned there is a school of breeders who adopt tactics greatly differing from any so far described. This can be termed the "guided" or "forcible" mating and its principles can be made applicable either to the table method or the floor method, and it is undertaken either single-handed or by two persons.

As before, when the animals are co-operative the natural "love-play" preliminaries should be indulged in for a little time. Breeders who practise the guided mating find that they get more satisfactory results by not giving the dog an opportunity to try to effect the mating himself, and at once assist him to negotiate and penetrate. The following is a description of this method. The passage having been examined by a sterilized and lubricated finger, and the pair having become acquainted, and the dog appears ready to serve, an assistant "A" lifts the bitch on to a table of good length but not too wide (approximately 4 ft. long by 2 ft. wide). "A" holds the bitch by grasping with each hand the loose skin on each side of the neck. If "A" presses his elbows to his sides, making the forearms rigid, the bitch is kept steady without exerting undue pressure on the neck. "B" then places the dog on the table, passing his left forearm underneath the bitch, with the flat hand, palm upwards, protruding on the far side. This prevents the bitch sitting when the dog mounts. The dog then mounts and "B" with his right hand carefully guides the dog's penis to the bitch's vulva, and actually inserts it. Insertion and penetration having been effected, "B" now immediately removes his right hand and places it, fingers downwards, flat against the dog's rump with the thumb round the base of the tail. "B" then gives him a gentle lift with the right hand and the dog places his hind feet on "B's" left hand and forearm, so obtaining a platform. By gripping the base of the tail the dog can be controlled if he is too vigorous. In due course, and conditions being propitious, the dog is turned in the usual way, both tails being grasped together in one hand.

Clearly this method cannot be employed with all breeds; it is one suitable mainly for the small and the short-legged varieties. It is favoured by terrier breeders of the Cairn, Scotty and Sealyham types.

The *principle* of the guided mating may be followed with any breed, with or without a table. Compromise and modifications become necessary as breeds increase in size and where the use of a table would be impossible. With the larger breeds, when single-handed a combination of the hitching-post (staple in wall) and the guided mating is a satisfactory and popular method.

The guided method has much to recommend it. Its devotees claim that much time is saved, and much useless expenditure of energy on the dog's part avoided. In the other methods it cannot be denied that dogs do fumble and jitter for a considerable time, until they make contact with the penis against the vulva. When contact is made, however, penetration is usually instantaneous. It is finding the vulva which seems to be the dog's biggest difficulty;

when found for him by the last method described the dog is quick to appreciate the contact and reacts rapidly. Fruitless efforts to find the vulva tend to make dogs impatient and erratic, and they will sometimes "spend" themselves outside the bitch as a result. The chief advantage of the method, then, is the saving of time and avoidance of waste of energy by the dog. The chief disadvantage is that many dogs resent being handled at all, and are completely put off if touched.

The opening remarks in this chapter showed that there is no one and only correct method of managing a service. But the novice, and perhaps more experienced breeders, should now be in a position to overcome any difficulty encountered in a service with any breed. If one method fails, the novice now has knowledge of alternatives. After a few cases, with fresh practical experience gained at each, he will be able to arrive at a routine suiting his individual requirements and the requirements of the breed of his choice, and more particularly the peculiarities of his own dogs.

CHAPTER V

AGAIN the reader can have a respite from the intricate
details of stud work in so far as they concern actual mating.

Stud work is a business and there is much to be learnt
regarding its financial side, as will be explained in this
chapter. If the advice is followed the novice will find this
aspect simplified for him, and at the same time the
guidance may enhance his reputation for probity as a stud
manager and minimize the possibility of a tarnished reputation resulting from a misdemeanour, committed through
ignorance. There are certain unwritten laws which have
come to be recognized as accepted customs and usages.
When obeyed they help to bring to a satisfactory conclusion any contract entered into; when departed from,
bad feeling is created, muddles and mistakes in pedigrees
occur, and in extreme cases litigation has to be resorted to,
or censure from the Kennel Club is invoked.

THE STUD FEE.

This is entirely a matter for the stud dog owner to fix.
The amount is extremely variable, being dependent on
the popularity of a breed, supply and demand in that
breed, the achievements and qualifications of the dog and,
finally, the quality of the progeny the dog throws. With
some breeds the fee is as low as ten pounds; in some cases
a stud fee of 260 pounds has been charged.

Taking a general average of all breeds, the fee for a
top-ranking dog, the quality of whose progeny is known
to be equal to that of his own, would be thirty to fifty

pounds. For good specimens, either champions or big prize winners, the average is twenty-five to fifty pounds. For lesser lights and dogs used perhaps only for pet litters, the average fee is ten to fifteen pounds. It is advisable for a novice with a good dog not to overcharge for its early services, or until its progeny has been proved; bitch owners are chary of using an unknown dog and in these days of financial stringency the dog might get no work if a large fee were asked. It is better gradually to increase the fee as the dog's potentialities warrant, and quite legitimate to increase it when the progeny is proved to be prize-winning stock, and the dog himself enhances his qualifications. As a simple example, if a really good dog in an average breed begins at twenty pounds it would be sound to charge thirty or forty pounds after each of the first two C.C.s are gained; on becoming a champion, from twenty to thirty pounds would be a fair increase. If the progeny was proving excellent from a variety of bitches, showing that the dog was prepotent, which means capable of *always* producing his own qualities from a large range of bitches, for this reason alone an increase up to thirty-five pounds is justifiable. If the dog becomes a multiple C.C. winner, and his services are much in demand, and the owner does not wish him to be overworked or abused, it is fair and advisable to limit the number of bitches served while at the same time maintaining his own deserved financial reward by increasing the fee to between sixty and eighty pounds.

APPROVED BITCHES.

It will be appropriate here to explain this term. It is a common fallacy among newcomers to imagine that it is always the stud dog who is responsible for the excellence of the progeny, and to think that because a dog is a marvellous champion embryo champions will always be sired by him out of inferior bitches. This is far from the truth. It is an established fact that some super-excellent dogs never sired a worth-while specimen in the whole course of their careers, which confirms that the real value of a stud dog lies in the quality of his progeny, not in the excellence of the dog himself. The bitch plays just as great, perhaps a greater, part in determining the number of the progeny and particularly the quality. This fact is the basis of another breeding dictum—that only specimens of the highest quality should be selected for mating if equally high quality is to be expected in their "get". The owner of an exceptionally good dog siring good stock is extremely unwise if he accepts for service all and sundry bitches of a mediocre or even inferior or bad calibre. If he does so it is inevitable that a large number of bad specimens, probably whole litters of nondescript stock, will be let loose, all bearing his dog's name as sire. The average breeder, ignoring the fact that the bitch dam is equally or more responsible, is quick to note the inferiority of the stock the stud dog is throwing, and it rapidly becomes published abroad that this quite good sire produces faulty progeny. Then the dog as a stud dog will be dropped like a hot penny. So is a stud dog's reputation damned. But an astute sire owner, with a stud dog of proved achievement and proved as an excellent stock getter out of suitable bitches with line-bred pedigrees, takes care not to have his dog's prospects so ruined. He is far-sighted enough to see that it is better for his Kennel account in the long run to have fewer stud fees rather than let his dog lose reputation. He ensures that bad stock shall never be discussed as having been sired by his dog, by being discriminating in the bitches he accepts for the dog to serve, and advertises the service to "approved bitches" only. He may want to see the pedigree of the bitch to make sure that the blood lines are compatible with those of his dog, and he will wish to see her to convince himself that her quality is such that only good stock can justifiably be expected. He reserves the right to refuse to accept for service an inferior bitch with glaring faults. Ultimately he is unquestionably a gainer and not a loser; the progeny start to win, and as no bad specimen has been mated by his dog the demand for its services will steadily increase, with an increasing stud fee, instead of income being cut off short as a result of evidence of bad stock. To go back to beginnings (after visions of twenty-guinea stud fees!), it is quite customary on introduction of a new dog to advertise him without fee for a service or two to "prove" him. This is fair, as the dog might be sterile, or impotent, and a fee would have been taken with no prospect of value in return. Until a dog is a proved sire perhaps the fairest to both parties is to take no fee at the time of service, but have a written agreement that the fee shall be paid subsequently if the bitch proves in-whelp. Services without fee for the first few times are also useful opportunities for the dog owner to train the dog with no financial worry if some services are abortive. The bitch owner takes the risk of missing a season but wins in the gamble if a litter results, its inception not having cost anything.

At any stage in a stud dog's career an arrangement is sometimes made whereby instead of paying a fee the owner of the bitch will hand over a whelp in lieu. This is generally to the advantage of the stud dog owner, since the selling value of the whelp is greater than the fee, and he has borne no cost of rearing. It is assumed that bitch owners ask for this concession only for personal reasons, in that they do not wish to part with cash. It cannot be too strongly emphasized that any such arrangement should be drawn up in writing, in duplicate to be signed by both parties and one copy retained by each. Gentlemen's agreements either between friends or strangers are well intentioned, but sometimes memories are short. Written evidence is always confirmative, and in the case of disagreement or non-compliance the written word is proof of the original intention. The document should cover all likely contingencies, so that there will later be no loophole for acrimonious argument. For example, in the case of a puppy in lieu of a service fee, agreement should be arrived at to cover the possibility of only one puppy being in the litter. Strictly, this should go to the stud dog owner; a small litter is an act of fate and bad luck for the bitch owner, but the stud dog owner cannot live on sentiment and must be paid for the service in cash or in kind. Again, if there are two or more puppies agreement should previously have been reached as to whom the first choice goes. If no litter results the stud dog owner has still to be paid for the service, but this is usually avoided by arranging that the same agreement shall be entered into at the bitch's next season. If the bitch whelps a litter the stud dog owner has fulfilled his part, but if all the whelps should die before the stud dog owner receives his pup the bitch owner has not fulfilled his. Here it should have been previously agreed that if for any reason no whelp is forthcoming in lieu of the fee either the fee or part of it shall subsequently be paid to the stud dog owner. Alternatively, the same conditions can be again imposed on the bitch owner, and a puppy shall be due from the subsequent season and litter. The above are only examples of things that should be guarded against. The suggested forms of agreement are not hard and fast rules; the actual terms are for the parties concerned to work out for themselves, but they MUST be previously agreed upon. Generally, a stud dog owner gains a good reputation, and increases his clients, if he tempers hard business instinct with clemency towards the bitch owner. Whatever the conditions, when once amicably decided upon they must be put in writing, and thereafter be rigidly observed. Written obligations ultimately fulfilled breed goodwill; haphazard oral agreement may cause ill will.

FOR WHAT IS THE STUD FEE PAYABLE?

On this point there is often among novices confusion leading to controversy, since some are under the impression that if no litter results no fee is payable, or that it should be returned. There is no doubt on this point whatever. If the dog is a proved and potent sire the fee is due, being payable for the service itself irrespective of the result. This is a hard and fast unwritten law. Moreover, the stud fee is due in advance before the mating. A stud dog owner having received a bitch by rail would be quite within his rights to refuse to carry out the service until paid. A stud dog owner is selling a service which, like any other commodity, has to be paid for in advance if ordered by mail, or paid for at the time if sold over the counter. One does not expect to buy something in a shop and not pay for it; similarly, stud dog owners cannot afford to run the risk of having to run prolonged credit accounts for stud work. It is unlikely that a stud dog owner would in actual fact stand on this right and leave a bitch to miss a season, but bitch owners should not place the stud dog owner in the position of having to contemplate such a course. However, if a stud dog owner has previously been "bilked" by an unprincipled bitch owner such experiences might harden him. It is obvious that once a bitch has been served, the "goods" have been sold; it is also obvious that they cannot be claimed back, because they have not been paid for. It therefore behoves the bitch owner to conform to custom and forward the stipulated fee before a mating is completed.

It is also customary that the stud dog owner shall not bear the expense of the bitch's return journey; the return fare should be prepaid and added to the stud fee. The foregoing is the only financial obligation of the bitch owner to the stud dog owner—but the dog owner has several obligations, general and financial, and if his reputation is to be maintained he should invariably follow the following routine:

A. If the bitch is brought to him:

(1) He should parade for inspection the selected stud dog as evidence that it is fit and well, and is the dog who is actually going to be used.

(2) Many stud dog owners prefer that the bitch owner (or person accompanying the bitch) shall not be present at the actual mating as strangers may put the dog off.

When the mating is effected, and a tie has resulted, the manager should then call the bitch owner in, as proof of a *bona fide* service. At this stage the animals are seldom put off; the bitch owner can remain, and by caressing the bitch increase her comfort and confidence.

(3) If no actual tie has been effected the stud dog owner has to make an honest decision himself by answering these questions: Was this mating adequate? Can a litter reasonably be expected? Am I justified in taking a fee? Remember, though, many bitches and dogs do not tie.

It has been mentioned elsewhere that the tie is not an essential, and that a tie lasting an hour has resulted in no litter while penetration for under a minute has produced a record one. In his very earliest experience of stud work the author was perturbed on this point, considering that a mating of extremely short duration could not result in a litter, and was in a quandary as to whether the fee should be returned. Experience has ended such doubts, but the novice requires some criterion to help him to make a decision. The following is suggested:

(*a*) If the dog has taken a dislike to the bitch (sometimes, though very occasionally, this occurs) and he appears to have no stimulation for her, he may make a perfunctory mounting and then show complete disinclination to serve and take no further interest. In such a case it is daylight robbery to accept a fee.

(*b*) The dog has mounted, appears to have penetrated, but *instantly slips away* and is functioning only outside, and if on further attempts the same thing happens no mating can have been effected and no fee should be accepted.

(*c*) If the dog has only very partially penetrated, and obviously all his exertions are futile and he cannot penetrate to any extent, this may indicate a stricture or obstruction in the vagina or a very narrow passage, or that the bitch has "dried up" or is completely "unready". This again indicates a useless mating; no litter can be reasonably expected and no fee should be taken. In these cases there will not have been any swelling of the bulb outside and without this a true service is extremely unlikely.

(d) The dog makes complete penetration, has visibly swollen outside at the same time, and when pressed to the bitch remains there, still swollen, for at least ten to twenty seconds and then comes out. This is a state of affairs quite different from the instantly slipping in and out situation. Here a mating can, and almost always does, result in a litter (provided that the bitch is fertile, though fertility or otherwise of the bitch does not affect the question of a legitimate service) and the owner may have complete confidence in accepting the fee. Without this information, he might have had qualms, feeling that the mating was too short. Thereafter while still in the "untied" position for any period, whether under a minute or remaining so (while held by the assistant) for ten, twenty or thirty minutes, a litter may be expected.

(4) Where there has been no "tie" the stud dog owner is unable to show the service as evidence. Having come to a decision about the adequacy or inadequacy of the mating, he should straightforwardly explain the circumstances to the bitch owner and if necessary not accept a fee. If the mating was very short, but in his opinion adequate, he should state that he is satisfied that one was effected and that a litter will result. The fee is then fully justified.

B. *If the bitch is sent by rail.* Except that the stud dog cannot be paraded, or the "tie" shown, the same procedure is followed, and details of the service should be sent either by 'phone or letter. Occasionally the stud dog owner is unable to effect a mating by the dog selected but considers one might be possible by another dog. He should at once communicate with the bitch owner, requesting permission to use another dog. It would be most unwise to use a different dog without first obtaining this permission.

The service having been satisfactorily effected, the stud dog owner must supply the bitch owner with the dog's pedigree or with its stud card. This enables the bitch owner to make out the pedigree of the whelps in the forthcoming litter. The stud dog owner must also supply a certificate of the mating and a receipt for the service fee. Mutual obligations are then ended.

THE STUD CARD.

When a dog is established at stud the owner will frequently be asked by prospective users for its particulars. The most convenient and time-saving way of providing this information is by having a stud card printed. The best form is one of cardboard, folded down the centre, thus giving four sides. The front side should be headed by the name of the Kennel, and date of its establishment. It is useful to reproduce a photograph of the dog, with the name above it in bold letters. An unfaked photograph is preferable; it should display the dog's natural points, not a photographer's deceptive skill. Under this, the owner's name and address are inserted, together with telephone number and nearest railway station to which bitches can be sent. On the reverse of the outer page, and on the left of the inside of the folder, a true copy of the pedigree is printed, also details of colour or markings, date of birth, its Kennel Club registration number, or Stud Book number. On the right side of the inside space is left for details of the dog's principal achievements and prize-winning record. On the fourth page, or outer half of the card, a description of the dog is given. This may state any points in the standard in which the dog excels, and any particular feature he is known always to transmit. A reference to temperament is advisable. Following this, the conditions of service should be stated, which include the fee, stating that it must be prepaid, plus cost of return carriage of the bitch. It should state here that though all possible care will be taken of visiting bitches they are accepted only at owners' risk. It is advisable to have this in print as it establishes that a condition of the contract is that the dog owner shall not be held responsible for loss or accident.

Finally, the terms relating to a free service in the event of the bitch missing should be stated.

THE STUD RECEIPT AND CERTIFICATE OF MATING.

This can be combined on one form, as in the following example, published by Our Dogs, Oxford Road Station Approach, Manchester:

STUD SERVICE RECEIPT

Date..19..........

Bitch ....

Owner ....

....

Stud Dog ....

Date of Service :—

      1st ....

      2nd ....

Due to Whelp ....

Received Stud Fee £   :

      Signature ....

Other arrangements (if any) :

....

Many commercial printers have similar forms already printed, which only require overprinting with the owner's Kennel name and address. Such Kennel stationery adds to a Kennel's reputation for efficiency. The forms should be detachable, leaving a counterfoil on which the owner can insert the details, providing a convenient record of all bitches served by his dog. The following is an example of the counterfoil:

```
┌─────────────────────────────────────┐
│                                       │
│  Date .............................. 19........  │
│  Bitch  ..............................................  │
│  Owner  ............................................  │
│  ..................................................  │
│  Stud Dog  ........................................  │
│  Date of Service :—                  │
│        1st  ......................................  │
│        2nd  .....................................  │
│  Due to Whelp  ..............................  │
│  Stud Fee £ ...................................  │
│  Other arrangements (if any) :       │
│  ..................................................  │
│  ..................................................  │
│                                       │
└─────────────────────────────────────┘
```

THE FREE SERVICE.

It is the accepted custom, and a gesture of generosity on the stud dog owner's part, to offer a free service if the bitch "misses", i.e. proves not to be in whelp. Usually a free service is offered to the bitch at her next season, but a proviso should be made that this is dependent on the dog still being in the owner's possession. Alternatively, a stud dog owner may offer a service from another dog or to any other bitch, in order that at some future date the bitch owner shall receive some value in return for the fee paid. These must be regarded as *ex gratia* gestures on the stud dog owner's part; he has fulfilled his part of the contract if a service has been adequately accomplished and it is not his responsibility if no litter results.

The stud dog owner should ask the owner of the bitch to inform him of the result of the mating. While it is courteous of the bitch owner always to do so it is essential, if the bitch has missed and a free service is claimed, that information of the failure to whelp should be given within seven days after the due date.

CHAPTER VI

Records of stud work.

HAPHAZARD methods in keeping notes or records of services effected by one or several stud dogs in a Kennel can cause chaos in the compilation of pedigrees.

It must be remembered that the word pedigree implies an accurate account of all the forbears of a dog or bitch; it is a genealogical table. One wrong entry or wrong record can destroy the value of a pedigree, possibly introducing a completely new ancestral line. Scientific breeders who base many of their breeding plans on "lines" decided upon after the study of a pedigree can have their work wrecked and their plans brought to nought by a single wrong entry. For example, a breeder might want a certain ancestral dog to appear in the pedigree of a dog he was studying and contemplated using, finds the ancestral dog is included and uses the dog. If, however, the inclusion was an erroneous one the breeder would fondly imagine he was "getting into" the wanted blood, though in actual fact it was absent. To give another example, a breeder might intend to practise a very sound policy of putting a bitch back to its grandfather, a champion dog, and on referring to his bitch's pedigree he finds the grandsire to use. If in fact the champion dog shown was not the actual grandsire, but entered on the pedigree in mistake for another dog, the bitch owner's breeding plan is abortive. The author knows of many cases where circumstances similar to the above have occurred through negligence or lack of method. Unfortunately, in dog breeding, as in any other walk of life or sport, there is a very small minority by whom shady methods are practised. There have been definite cases of intentionally faked pedigrees, forged in order to give progeny a high value. The Kennel Club has its own regulations for dealing with such cases and the subject is beyond the scope of this book. An attempt will be made, however, to explain to the novice how, by accurate record keeping, he can avoid errors which might involve him in Kennel Club censure.

The Irish Kennel Club has an excellent system for keeping pedigrees accurate. When a dog is registered with them they write to the owner of the stud dog sire, stating that his stud dog is recorded as the sire of the animal to be registered, and requesting verification from the stud dog owner that the mating took place. Our own Kennel Club's present application form for the registration of a dog is a vast improvement on the pre-war one as it greatly tightens up the authenticity of a pedigree, space now being provided for the signature or name and address of the owner of the stud dog. In either case, however, the information may be requested months, perhaps years, after the actual mating took place and no breeder can be expected to rely on his memory for an accurate confirmation. A simple recording system removes the strain on memory and ensures efficiency and accuracy. Again, inaccurate records in Show entry forms can cause mistakes and disqualification. A person buys a dog, enters the pedigree details of sire and dam on the entry form, and is ultimately disqualified, the Kennel Club informing him that there is no record of his dog having been sired as stated.

The date the service took place is always important; it affects the age of the whelps, which affects their eligibility for entry in certain Show classes.

Many other examples of the penalties of inaccuracy could be quoted, but the foregoing should suffice to make out a case for the necessity of record keeping.

The following are considered to be the essential details[1] of which record should be made:

[1] In a previous work of the author (in collaboration with Flight-Lieut. P. Townshend), "*My Dogs*" *Kennel Records Book*, space is provided for the accurate recording of every essential detail connected with stud work.

Each stud dog in a Kennel should have its own page or Record Book headed by his name, his sire and dam, his Kennel Club registration or Stud Book number. The page should be divided into columns, or spaces provided for the following headings:

Column 1. *No. of the service.* This is useful in indicating the extent to which a dog has been used, to make comparisons with the other stud dogs, and as a means of quickly ascertaining the financial earnings of the dog.

Column 2. *Date of the service.* Important when any query is raised, either by a breeder or by a Kennel Club, requiring verification of a service or pedigree.

Column 3. *Name of bitch served.* Essential for verification purposes.

Column 4. *Name and address of owner of the bitch, with telephone number.* Necessary if any further correspondence is needed in connection with the service. It is also useful to the owner if he receives frequent offers to buy his stock. If he has no stock of his own available he can either refer the enquiry direct to the bitch owner or undertake to "buy in" some of the progeny from his own dog, and thus have stock available for resale. Here it is useful to note also the quality of the bitch served, as a guide to whether the progeny would be likely to be up to standard.

Column 5. *Date the bitch is due to whelp.* Considered in conjunction with (4), this indicates when progeny from the dog would be available, and also the age of the pups.

Column 6. *Details of the service.* These are useful to record as evidence of a possible reason if the bitch proved not to be in whelp. Brief notes should be made on the following points:

   (*a*) The number of days the bitch has been in season.
   (*b*) Whether there was a tie, and its duration.
   (*c*) If no tie the actual time penetration lasted.
   (*d*) Whether the service was straightforward, or whether the bitch was very difficult to serve, being nervous and unco-operative.

(e) Any circumstance which might make conception doubtful.

Column 7. *Result of the mating*. Entered when informed by the owner of the bitch of the details of the litter. It should include the total number of whelps, the number of dog pups and number of bitch pups. If it is a breed in which there are many colour variations, the colours of the whelps should be included. These records are useful for the owner subsequently to gauge the potency of his various stud dogs, the proportion of dogs to bitches thrown, and whether a certain dog throws a preponderance of any particular colour.

Column 8. *The date the stud fee was paid and the amount.* Useful in assessing the stud earnings of each dog.

Column 9. *Conditions*. Finally, space should be provided for a note of the conditions of the service, as for example:

(a) Whether for a fee. (b) Whether for a pup in lieu of the service fee. (c) Any other special condition. (d) If information is subsequently received that the bitch "missed" it must be recorded that a free service was promised, also of the terms of the free service.

A stud dog owner with many stud dogs serving many bitches in the course of a month will find it difficult to remember exactly the conditions and results of every service, but notes like these will enable him to conduct his stud work accurately and so eliminate any possibility of dispute later.

A final word of advice to the novice stud dog owner; it relates to his advertisements. He may have a very good dog, but the wording of his advertisements perhaps does harm to his prospects because so often it discloses ignorance and inexperience. One can best explain this by examples. One frequently sees a dog advertised as having been "sired by a bitch" and "out of a dog" by the wrong use of the words "ex" and "by". The word "ex" means "out of", and therefore precedes a dam's name, while the word "by" precedes a sire's name, indicating that it was sired by that dog. To see a stud dog advertised as "by" Jane "ex" Bob, though amusing, destroys confidence. Also, the term "line-bred" is sometimes inappropriately used. As it implies a breeding-down for many generations it can be applied only to a dog of good age, or even one of a past decade, yet one sees stud dogs described as "line-bred" to a very recent champion probably only eighteen months old, who could not have had sufficient descendants to make "line-breeding" from him possible. Further, one perhaps sees a stud dog described as containing the blood of possibly half a dozen leading strains, with names of widely divergent types of champions, quoted by an advertiser presumably under the impression that the greater the number of champions from diverse blood the better the dog must be. To a student of pedigrees this would create a mental picture of an animal resembling a patchwork quilt, rather than of one with a pedigree built up on a scientific plan. Grossly exaggerated advertisements with super-eulogistic phrases cut no ice with the knowledgeable breeder. The novice is therefore advised to stick to true facts correctly stated in moderate language, which are more convincing than flamboyance and palpable inaccuracy.

PART II

Whelping

PREFACE

SOME of the subject matter of Part II of this book was dealt with in a series of articles in *The Dachshund*, the magazine of the Dachshund Club. The advice given has been acclaimed by novices and experienced breeders alike as being of real value and practical help.

The gist of many letters and expressions of appreciation was that since perusal whelping cases were approached with confidence instead of trepidation.

It was suggested to the writer that the subject should be given greater scope, and that a large number of breeders would benefit if the notes were published in book form and made to cover all breeds. Consequently the articles have been completely revised to make the advice more generally applicable.

It is obvious that with over ninety different breeds of dogs there will be whelping factors and eccentricities peculiar to each breed. At the same time, the fundamental, natural biological phenomenon of whelping is constant for all breeds. These general principles of whelping should be known and understood by all owners of bitches of any breed if it is intended that litters are to be bred. Such knowledge will enable them to co-operate intelligently with the veterinary surgeon, particularly when circumstances make the employment of his services essential. Also, the knowledge will help owners to act on their own initiative when difficulties are met with, and when veterinary help is not available. The author has purposely reduced to a minimum theories and technicalities in order that the advice given shall be of a practical nature, based on personal experience over a period of many years. It is confidently prophesied that the lives of many bitches and their whelps will be saved, and much suffering spared them, if this book is studied and its simple, straightforward guidance followed.

In order that the veterinary profession shall not regard the advice given in this book as an infringement of professional rights it is reiterated that its main theme is co-operation with the veterinary surgeon at all times. If some of the advice is a little more advanced than simple first aid, it is given because in many emergencies a veterinary surgeon's skilled treatment is not immediately available.

The writer hopes that no veterinary surgeon will resent information and advice being given to breeders which, in certain circumstances, will save a bitch from suffering, or possibly its life and the lives of its whelps, when a veterinary surgeon is not able to be present.

A veterinary surgeon is a busy man and cannot be expected to stay throughout a whelping case lasting twelve hours, and it frequently happens that an emergency will arise shortly after he has ended his first visit. There are outlying districts where there is only one veterinary surgeon to cover a very large area; owing to distance and other calls it is conceivable that half a day might elapse before he could attend an emergency. Further, we have the winter litters—intense fog, ice-bound roads, transport at a complete standstill—when the attendance of the veterinary surgeon is impossible. In any of these circumstances the writer considers that the veterinary profession would prefer that owners should have enough knowledge, combined with confidence, to act promptly rather than wait until too late, or through ignorance find themselves unable to relieve a minor complication, which might otherwise cause suffering.

Again, in Australia and New Zealand there are many outlying districts so distant from a large town that the inevitable delay in a veterinary surgeon's attendance can be not merely a matter of hours but days. In these areas there are many keen breeders to whom this book should,

I hope, prove of value, since the knowledge they will gain will enable them to act on their own initiative when there is *no* prospect of obtaining professional help. The author knows of definite cases of litters and pups being lost in certain parts of the Commonwealth owing to unavoidable lack of skilled assistance. In many of these fatalities, had the owner practical knowledge and confidence, the tragic losses could have been avoided, a bitch's life saved and a profitable litter reared.

In human childbirth it is by no means rare for delivery to take place before the arrival of the doctor. It is accomplished by the midwife, who has enough knowledge at least to carry on. With a whelping in similar circumstances it is the owner who must turn midwife, and the more knowledge and confidence he possesses the better will be the result.

Advice in this book is intended primarily for the newcomer, though procedure with certain of the more advanced cases of abnormality will prove equally instructive to the experienced.

Today there is a vastly increased number of people who have only recently taken up dog breeding as a hobby or a business. Most of these are very enthusiastic, but if one may be permitted to say so, many are woefully ignorant about whelping. This assertion is based on the many S O S 'phone calls one has received at whelping times, and on the innumerable questions asked on the subject.

It is to avoid unnecessary suffering to bitches and their whelps, and disappointments and financial loss to their owners, that it has been decided to put experience into print, in the hope that this may help readers. Care has been taken to ensure that the information given is authentic, that the veterinary aspects are fundamentally accurate, and that no advice given could possibly be harmful.

Descriptions of complications and somewhat lurid

details of types of whelping cases are given, to indicate the sort of difficulties that may be met with, so that the inexperienced may know what to expect and thereby avoid being taken by surprise. The advice is based entirely on personal practical experience gained by implicitly carrying out a veterinary surgeon's instructions, and on conclusions formed from whelping cases of several different breeds, but chiefly Dachshunds, over a period of twenty-five years.

CHAPTER VII

Further reference to practical application of anatomical detail, relating to the bitch only—Principles of whelping—Co-operation with the veterinary surgeon at whelping time.

In Part I, The Service, the reason given for describing in detail certain anatomical structures was the value of the practical application of this knowledge. In just the same way anatomical knowledge of the genitals of the bitch will prove beneficial when applied practically in whelping cases.

(1) *The Vulva.* An early sign that actual labour will not long be delayed is that the vulva enlarges considerably, becomes very soft and pliable, which facilitates expansion, and at the same time a discharge is observable.

(2) *The Angle of the Vagina* from the vulva. Any breeder may expect to have to assist a bitch to deliver a breech presentation (see relevant chapter and illustrations). It must be emphasized that the "pull" is in a *downwards* and outwards direction approximately at an angle of 30° to the rump. Any traction in a horizontal direction, or at right angles to the rump, could injure and cause pain to the bitch and whelp, and make the latter's exit more difficult.

(3) *The Length of the Passage.* (Described in Chapter I, The Service.) The distance from the vulva to the brim of the pelvis or beginning of the body of the uterus; the lumen or width or diameter of the passage; the presence or otherwise of adhesions, are all of the utmost importance. It is explained in a later chapter that it may become imperative to make a digital examination of the passage for the following reasons: (*a*) In order to find out if a whelp has passed over the brim of the pelvis. (*b*) To ascertain what complication is causing an obstruction or delay. (*c*) To determine how far down the passage

a whelp actually is. (*d*) To turn the feet of a breech presentation. In any such circumstance the anatomical information given in Part I will assist practical manipulation. Here it must once again be stressed that although it is doubtful if any veterinary surgeon would object to a digital exploration of the vagina as a preliminary to a service, the veterinary profession is averse to any such amateur interference immediately before, or during, an actual whelping. Therefore a novice should not do this until he has gained experience by previously watching the procedure, or does so only in the presence of a veterinary surgeon whom he has asked to instruct him in the matter, in anticipation of an emergency when alone. An experienced breeder can, and often *must*, do so. But in either case the antiseptic precautions mentioned in Chapter I, The Service, must be taken. Again, in either case such digital examination or assistance should be done only when, owing to "time lag" or other emergency, intervention is necessary and there is no hope of a veterinary surgeon arriving in time, and the owner feels he must do something on his own initiative.

(4) The peculiar arrangement of the uterine horns running along each side of the body indicates during the last two or three weeks of pregnancy where to make an *external* examination by feeling for the presence and movements of whelps; with experience it is often possible to determine approximately the number of whelps to be expected.

(5) The description of the position of the uterine horns assists, when one *thinks* that the whelping is all over. It is quite impossible for a veterinary surgeon to make an *internal* digital or forceps exploration for a whelp left high up in one of the horns, but by knowing where and how to feel externally the presence of another whelp may be detected when otherwise one might imagine that the bitch had settled down comfortably with her labour trials and tribulations all happily ended. This possibility

of an additional whelp, or whelps, is very important, as will be seen from subsequent pages. This is mentioned specifically because so many novices feel for whelps only quite low down in the abdomen, and not much beyond the groin, not realizing the peculiarity of the extensive uterine horns. This knowledge will enable them to ascertain, by external palpation, the presence of whelps when otherwise they would have had no idea where to feel.

Considerable amplification of the above headings will be found in subsequent chapters. They are briefly men-mentioned here to recapitulate earlier anatomical explanations, and to make more intelligible references to structures.

Further, in the later pages there may crop up odd items of advice which may appear redundant, having already been dealt with in Part I, where it was expedient to discuss them. Many were sub-headings of subjects dealing with breeding generally, and could not be discussed in relation to *either* the service or whelping only. An occasional repetition in Part II therefore may be beneficial, on the principle that "if one throws sufficient mud at a wall some of it may stick".

PRINCIPLES OF WHELPING. CO-OPERATION WITH THE VETERINARY SURGEON AT WHELPING TIME.

No two bitches behave exactly alike at whelping time. It is therefore impossible to lay down hard and fast rules about what may be expected before, during and after the appearance of a litter. Generally speaking, however, the same bitch will behave at each of her whelpings with the same individual mannerisms. These should be noted as they help to determine whether a bitch is a "good whelper and good mother" or otherwise. From this broad statement minor exceptions must be made, however. Firstly, although a bitch is "a good whelper" in her behaviour, and has had two litters with the utmost ease, a complication might arise in her third litter which could result in this whelping being a difficult one. Secondly, a very excitable bitch at her first litter will often improve at subsequent ones; while a bitch who had great trouble in producing on the first occasion may whelp easily afterwards though still showing her individual mannerisms. A lazy bitch who has no grit and will stand no pain never seems to improve and usually becomes worse and exhibits "uterine inertia" (described later) to the maximum extent. A carefully kept record of the details of each whelping will prove invaluable for future reference, and such record-making is strongly advised.[1] Details noted in these records should include:

(1) The number of days before or after the due date that whelping took place.
(2) How many minutes, or hours, elapsed between the arrival of each whelp.
(3) If delivery was normal or required veterinary intervention.
(4) Whether a good or bad whelper, with reasons.

No bitch should be bred from more than twice in three successive heats. Every other season is more humane, reduces the strain on the bitch and may be regarded as ideal. Never hesitate to call in a veterinary surgeon or grudge the fee he demands in a case of whelping. The prospective pups are valuable, as is your bitch, sentimentally and financially. One whelp saved by calling in expert veterinary aid when it is essential may pay the bill at least fifteen times over!

Advice must be given about the age when a bitch should be first mated. If a bitch comes into season very early, at from eight to ten months, it is extremely unwise to mate her at this immature age. It is much sounder to

[1] See "*My Dogs*" *Kennel Records Book*, compiled and published by the author and Flight-Lieut. P. Townshend, in which space for this data is provided.

wait till the second season, or until the bitch is turned eighteen months of age. The only exception is in the case of miniature breeds, when it is preferable to mate at an early age.

Before a maiden bitch is mated ask your veterinary surgeon to examine her. He may find certain conditions which would make a successful mating impossible, or some abnormality which would endanger the bitch's life at whelping time. Among such conditions are a very narrow vaginal passage, so small as not to allow the passage of the little finger. There may be an obstruction, growth, adhesion or stricture in the passage, which would prevent penetration by the dog. The pelvic bones may be so placed that the passing of even a very small whelp is impossible. In the latter case a veterinary surgeon may recommend a mating but advise against normal whelping and suggest a Caesarean section. This is quite safe, does no harm to the bitch, and a few hours after it she may be happily suckling her litter. She can be mated again if desired. Generally he will find nothing amiss and the mating may be carried out with the satisfaction of knowing that in all probability the whelping will be normal. Had you not arranged this examination your bitch might have had some unsuspected serious abnormality, which would have endangered her life had she been mated and allowed to whelp.

The bitch's normal gestation period is sixty-three days. Experience has shown that, in the majority of cases, bitches will go to full term. There is, however, nothing untoward if whelping occurs one, two or three days previously. If a bitch whelps on or before the fifty-eighth day something has gone wrong and generally the pups are too weak and immature to live. There is nothing one can do about this. Earlier miscarriage is extremely rare. It is usually due to some severe shock or violent accident. Should your bitch have a miscarriage it is imperative that you have her examined afterwards by a veterinary surgeon—this is just as important as his attendance during whelping. There is no cause for alarm if a bitch goes one, two, or even three days over her time, but beyond the sixty-sixth day matters cannot be quite right; it may be evidence of some abnormality, and a veterinary surgeon's attendance is advisable. Forceps delivery, or Caesarean section, may be necessary.

The following authentic case is given as an extreme example. A bitch (not the author's) was allowed to go ten days beyond her due date. The owner had been told by another breeder (presumed to be experienced): "Don't worry; she will be all right." Result: death of bitch and whelp. Cause: one whelp only, but an enormous one and impossibly big for a bitch of the breed to evacuate. Consequently, during these protracted ten days the whelp had died *in utero*, general septicaemia automatically super-vened, with the fatal result. The onus for this fatality must be put on the owner. The correct procedure after three days' delay was to have sought veterinary advice, and asked for an X-ray to be taken. The condition would then have been obvious; the large whelp, probably then alive, could have been safely removed by Caesarean section, reared and possibly ultimately become a C.C. winner, while the bitch would have lived to produce subsequent profitable litters.

Single large whelps are quite common in old bitches. This fact should help the owner to know what may be expected with them, and underlines the added precaution to be taken in such cases.

A number of other types of abnormality will cause a bitch not to whelp at full term, or within two days of it. Without attention, many could cause the death of the bitch, but with assistance its life could be saved. The moral: never allow your bitch to go more than three days over time without ascertaining from your veterinary surgeon the cause of the delay.

CHAPTER VIII

The phenomena of normal whelping in detail—Signs that whelping
is imminent—The course of an uncomplicated whelping.

SUCCEEDING chapters deal principally with difficulties,
abnormalities, and pathological conditions and complications at whelping time. It is quite impossible to understand these without a thorough knowledge of the
fundamental processes of a perfectly simple, straightforward, normal whelping. Moreover, as some later
details are a trifle unpleasant it is intended that this
chapter shall be devoted only to the pleasant part of our
hobby—a normal whelping. It is doubtless the gratifying
culmination of breeding plans made a year or two
in advance; of the service satisfactorily concluded; of
several weeks of care and affection from the time your
bitch proved "in whelp". It is the time when the
bitch and her owner are closer together than at any
other, and a time when she trusts and has to rely on you
implicitly. By confidence and *knowledge* let this be a happy
climax for both. Don't wreck your plans and aspirations, or
"let your bitch down" through carelessness and ignorance,
should the expected normal prove to be abnormal.

REACTIONS OF THE BITCH THE DAY BEFORE SHE WHELPS.
Assuming that the mating has been satisfactory, the
bitch is safely in whelp, has been properly exercised and
her diet increased and made more nourishing during the
last three weeks, she should be examined by a veterinary
surgeon three days before she is due. He will probably
be able to tell you if she is likely to whelp prematurely,
in twenty-four hours, or carry on to her full term. This is
important. A veterinary surgeon's time is too valuable

for him to spend a whole day or night with you; you must therefore yourself know and watch for signs and symptoms and be able to inform him of them. You should be able to recognize any situation in which he is wanted immediately. Maiden bitches vary considerably in their whelping behaviour. Some go to full term and whelp easily. Many will whimper, tear their blankets, scratch feverishly and "make their beds", or rush wildly from room to room one or two days before their litters arrive. Some try to find a secret place for their "lying-in". You must be prepared for all these tricks. Don't be misled by such antics into believing that the bitch is about to whelp at once. This behaviour may go on for hours or the whole day previous to parturition. It is sound at this stage to sponge the belly, nipples, vulva and anus with Dettol or T.C.P. diluted with warm water, and repeat whenever the bitch has gone outside her quarters. The incidence of worms in the whelps may be reduced by this procedure, and it avoids contamination of their first milk meal.

SIGNS THAT WHELPING IS IMMINENT.

One of the earliest indications that labour may be anticipated within 24 to 48 hours is a drop in the bitch's temperature; thereafter the following signs should be looked for and noted; they are all pointers that actual whelping will not be long delayed:

(1) The bitch settles down rather quietly, often stretched full out with head between paws.
(2) She pants and breathes heavily, perhaps giving a sharp cry as though in pain; these are the labour pains.
(3) She makes apprehensive turns of her head, looking anxiously at her rear parts.
(4) She will refuse food.
(5) She will more often than not vomit.

(6) The vulva is swollen and softens, and there is a clear mucous discharge.

(7) She will press hard with her rear against her box and make heaving or straining motions.

No. 7 is the most important sign of all, and at the first sign of straining you should look at your watch. The time elapsing between these first "heaves" and the delivery, or more specifically the non-appearance of a whelp, tells you whether the whelping is going to be quite normal or whether you may expect difficulty. It is the most important phase in intelligent co-operation with the veterinary surgeon if a complication ensues, and the information which will help him most. Invariably the veterinary surgeon's first question is: "When did she actually start in labour?" If you cannot accurately inform him he has to guess. If you can tell him "three, four or five hours ago", he can more readily decide what action to take.

THE BIOLOGICAL FACTORS, AND NORMAL COURSE OF AN UNCOMPLICATED WHELPING.

Labour straining having started, the intermittent "heaves" may continue for anything from five minutes to one and a half hours before a pup is produced. These should be followed by the appearance of the "water-bag" (a lay expression; the following is a technical description). The "water-bag" should not be mistaken for a whelp, as is so frequently done by novices (see Fig. 1). It resembles a mass of greenish-black fluid contained in a membrane or skin; it varies in size according to the time taken to pass it. Each foetal whelp is encased in a membrane, and contained in a sac of fluid. This is designed by nature to protect the whelp while in the uterus or womb from external shock or injury as it acts like a "buffer" or cushion. A further purpose of this amniotic fluid sac is that instead of the whelp's body emerging first, the uterine contrac-

FIG. 1

The water-bag presenting

tions squeeze the fluid into a "bag-shape", which by preceding the whelp facilitates its egress, having gently dilated the vaginal passage.

The way for the whelp thus prepared, the water-bag emerges and bursts, and a whelp *should* shortly afterwards appear, nose first with front legs facing forward and tucked under the chin. One vigorous labour strain will usually expel the whelp immediately and entirely.

The whelp is still deriving nourishment and oxygen via the umbilical cord, which attaches it to the placenta. The placenta, commonly known as the "after-birth", resembles a mass of raw meat, and varies in size from two inches in diameter to five inches according to the breed.

Cord and placenta should both come away with the whelp (see Fig. 2). The whelp being completely encased in the membrane, the normal bitch's instinct and immediate duty is to tear away this membrane with her teeth, thus allowing the whelp to breathe air for the first time in the post-uterine state. Next, the bitch should start to sever with her teeth the umbilical cord and finally devour the after-birth. This is a normal function; it provides the bitch with some nourishment, so that she requires no solid food during the first forty-eight hours. She should next energetically lick the pup and generally "rough house" it all over the whelping bed; this is to dry and warm it and to stimulate the heart action and lungs. By this time the whelp should have given its first tiny cry; when that happens one knows all is well, and in a few minutes it should fumble towards a nipple and start guzzling. The bitch heaves a sigh, settles her head between her paws, and awaits the next onset of pains, which should occur within half an hour and should never be delayed more than three hours. The whole process is repeated again and is continued until all the whelps have arrived, when the bitch settles herself down. The breathing is calm, deep and slow again; the anxious, strained look disappears. She takes on a seraphic expression

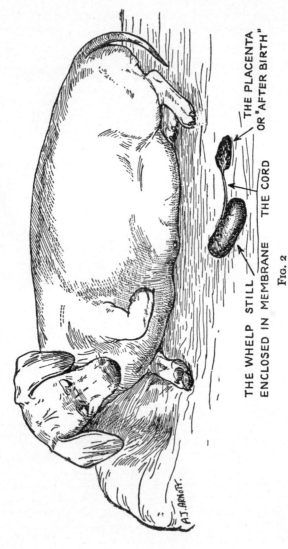

THE PLACENTA OR "AFTER BIRTH"

THE CORD

THE WHELP STILL ENCLOSED IN MEMBRANE

FIG. 2

A whelp and after-birth delivered

and has no further use for her master or mistress for at least two days, merely tolerating their presence, but violently resents intrusion by anyone else. A curtain should be put over the box. Keep her quiet and in semi-darkness, and withstand the temptation to show her puppies to all and sundry. She should be given a drink of hot milk, with glucose added, no solid food for forty-eight hours, and only milk, gruel, gravy and eggs.

A good plan is to weigh each puppy as it comes.

It is not always easy to ascertain whether the last or all the puppies have arrived, but by feeling the flanks a foetal movement may be detected if a pup is left high up in one of the horns of the uterus. Consequently, it should be made an invariable practice to let the veterinary surgeon examine the bitch the day following the whelping. This is, firstly, to ensure that no whelp has been left unborn (which can cause dire complications when one had thought all was well), and, secondly, he will give an injection to clear the uterus of any after-birth or membrane debris remaining. This reduces the risk of temperature from a septic condition and ultimate septicaemia.

Note that throughout the whole of this normal whelping the owner should merely sit beside the bitch; his presence and occasional soft-toned encouragement will give her all the assistance she requires. In a normal whelping as above described let your policy be: minimum amount of interference or assistance; know what she should do and act only if she fails to do it; know what complications to expect and act correctly when they happen. If the bitch behaves normally, and nothing untoward occurs, leave her alone; don't agitate her. It is important not to display anxiety and fuss too much over her during the forty-eight hours before the "due" date. Animals can sense such concern and react to it.

Later chapters will give you the knowledge you require, and help you to decide whether the whelping is normal or complicated, or whether your interference is required.

CHAPTER IX

General remarks on abnormal whelpings—How to minimize possible
   complications—Some statistics.

An abnormal whelping may be expected if from the
first "heave" three hours elapse with no sign of a birth.
Here a veterinary surgeon's immediate attendance is
strongly advised (having previously warned him to be
available if required). It is not meant that a lapse of
three hours necessarily indicates disaster, but it might,
therefore no time should be lost in securing expert assistance. If just before the veterinary surgeon arrives a
beautiful pup does pop out comfortably, so much the
better, but if there is a serious complication, such as a
lateral deviation of the head at the brim of the pelvis,
how glad you will be that the veterinary surgeon is on his
way to deal with it! He will remove the first whelp. Your
bitch is still fresh and strong—ready to settle down to
deliver the next. If, on the other hand, the bitch is left
unaided she will strain repeatedly, trying hard, but will
so tire herself that her muscles will become exhausted and
she will either "pack up" because she is lazy and will not
exert herself further, or because from sheer pain she is
incapable of further effort. The result may be the death
of all the whelps and possibly of the bitch herself. Early
removal of the cause of the trouble is essential; whether
the offending pup be delivered alive or dead, you may
save all the others and preserve your bitch's life.

It is emphasized that this "over three hours' lapse"
refers not merely to the arrival of the first whelp; it is
equally applicable after the delivery of *each* whelp. As
each whelp is delivered look at your watch and make a
mental note: "If the next water-bag does not arrive in

three hours from now I will call in the vet." Perhaps this is even more important than with the arrival of the first. Let us study a simple example. The bitch is perfectly fit; she has strained for fifteen minutes and pup number one pops out, suckles, and the bitch lies contentedly. Half an hour passes and a few heaves produce another lovely pup. Similarly, after a further half-hour pup three is delivered. You are now elatedly gloating over their perfection and have mentally assessed their value and even their disposal. The bitch, however, by her behaviour obviously has at least one or two more whelps to come. She is restless, seems a little in pain, is rather tired now, and although she has made some straining efforts for the last two hours she will not exert herself to the full and appears not to relish the thought of producing this one, and looks as if she is going to "pack up". A further hour has gone by but nothing has happened. This may be all right; on the other hand, it may be all wrong. Don't take a chance on it; ring your veterinary surgeon. It may be two or more hours before he comes, during which time your bitch's chances are decreasing if she has a major complication, or a minor one beyond your skill. When he arrives he can end the complication, or take whatever course he deems advisable. Your bitch is still fairly strong for the remainder, the earlier arrivals will not be orphaned and the following ones may be whelped simply. Take the converse picture: if you go to bed when only three whelps have been delivered, feeling "she will have had the rest by the morning", you may find her dead, her lovely pups weakening from lack of her care. Then all your financial castles crumble.

Before dealing in detail with minor and major complications, their recognition and treatment, it will be expedient to give some statistics and a classification ranging from normal whelpings to the most severe and fatal issues.

Study of at least a thousand cases would be

necessary before a completely reliable analysis could be made. The writer actually did appeal to other breeders to send him reports, on the lines of the classification below, of their whelping cases. From these he intended to collect and collate sufficient data to make possible a summary from which could be accurately gauged the percentage of normal to abnormal whelpings, with the intermediate degrees. The response was so poor, however, that he has had to rely largely on his own experience, which, with the few others reported, covers approximately a hundred whelping cases. Therefore no claim is made that the following statements are indisputable; they should be regarded as a rough guide only.

Normal whelpings; bitches and all whelps healthy and all survived ..	50 per cent
Whelpings with minor complications but bitch and all whelps survived ..	27 ,,
Whelpings with major complications, where bitch survived but some or all whelps were lost	20 ,,
Whelpings with major complications where bitch died and some or all of the whelps died	3 ,,

From this analysis two facts stand out. Firstly, that in one of every four litters there will be some complication causing the loss or death of whelps; secondly, that one bitch will die in approximately every thirty whelpings.

It is the first fact that so impressed the writer that it induced him to publish this book, in the hope that breeders, by gaining a deeper knowledge of the intricacies of whelping, can reduce the incidence of whelp mortality. On the other hand, the bitch mortality of 3 per cent can be regarded as low and encouraging, but the significant fact here is that at practically all the whelpings on which this analysis is based SOMEONE SAT IN ATTEN-

DANCE FROM START TO FINISH, AND THROUGHOUT EACH THERE WAS INTELLIGENT CO-OPERATION WITH THE VETERINARY SURGEON WHEN NECESSARY. One is convinced beyond a shadow of doubt that had this not been done in many of the 25 per cent cases of major complications more bitches would have been lost, and the bitch mortality rate would have been much higher. The so-called "natural" method of whelping can fairly be described as nothing short of cruelty—where the bitch is just left in some outhouse to "get on with it" and only on the next morning visited, possibly to find one or two "born dead" puppies (and probably unnecessarily so) or the bitch in dire distress and danger. The dog is no longer a wild animal living in the "natural" state when it preferred to whelp alone. It is a domesticated animal, and its accouchement calls for unstinted care. This does not mean coddling, over-interference or over-fussing. In many whelpings the writer has been connected with the bitch would not attempt to whelp alone, and became frenzied if left, obviously beseeching someone to "see her through".

So intense is the affection of a bitch for her master at whelping time that situations can become embarrassing and unmentionable in polite or "non-doggy" circles. The author's most typical experience of this was an occasion when a camp bed had been placed in the whelping shed on which to relax while waiting and watching. Whenever he left the shed the bitch screamed as if demented. In the small hours the bitch jumped on his chest and, with no due warning, there deposited her first whelp. A messy business, but if one loves one's dogs a lot must be forgiven. Having cleaned up the whelp, it was returned with its mother to the whelping box where both remained contentedly for an hour. She then, with the whelp in her mouth, again jumped on his chest, pushed the whelp against his hand with her nose, and with an expression of "you look after this" proceeded to deposit another whelp in exactly the same place. Having resignedly changed his attire for the second time he wondered how he could obviate a repetition. He couldn't. This procedure occurred five times; before each birth the bitch leaped from her box, each time with a whelp in her mouth and each time pushing it towards a hand, and delivering the next whelp on the one spot she had definitely decided was the only right one. Such is the faith, persistence, affection and obstinacy of the Dachshund. The foregoing is not fictitious narrative; it was, unfortunately for the author, only too true.

On another occasion while sitting in front of the kitchen fire for a winter whelping, another bitch could not be induced to deposit her whelps anywhere else than on the author's lap as each one arrived. Other breeders can recount similar experiences of this devotion and trust.

CHAPTER X

IN THIS chapter complications are detailed and advice given about action in emergencies. Step by step, we will consider MINOR complications at each stage of normal whelpings.

(1) A bitch is lazy and nervous; feeling pain and having made a tentative "heave", she cries and will make no further labour effort, knowing it will hurt. She will lie passively, sulk, and "pack it up". This has been known to go on for twelve hours and calls for veterinary action; an injection has to be given to contract the uterus and expel the whelp, or the whelps must be removed with forceps. Technically, this is an example of "uterine inertia".

(2) The "water-bag" appears, bursts, but is not followed by a whelp. If a period of more than two hours elapses some complication must be feared, especially if the bitch continues to strain heavily. A veterinary surgeon must always be contacted after this period.

(3) The bitch, particularly a maiden one, when the first whelp appears either has no instinct to attend to her immediate duties or is too frightened to do so. Here you must assist. It is essential that the membrane covering the whelp is removed to allow it to breathe. Pull away the membrane from the head and strip it from the whelp. Also separate the whelp from the placenta—that is, sever the cord. You must take care to avoid one of the chief causes of an umbilical hernia. An umbilical hernia is a protrusion of the gut through the abdominal wall at the

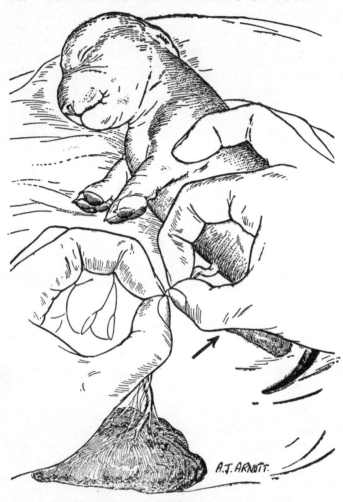

FIG. 3
The severance of the umbilical cord. Arrow shows
the right hand pulling away from the left

region where the cord from the placenta is attached to the whelp. Faulty manipulation here will put a strain on the abdominal muscles and cause a hernia. (See Fig. 3). The correct way to ensure that this does not occur is: Pinch the cord in the thumb and forefinger of the right hand about one inch from the whelp's body, pinch the cord between the thumb and forefinger of the left hand, with the thumbs of each hand touching. Pull with your RIGHT hand, i.e. pull the pup away from your left hand; NEVER pull the cord from the whelp's body. The average cord will sever after a twist and pull; if it is unduly tough, use your left hand to cut it with a pair of sterilized scissors close to your right thumb. An alternative method is to ligature the cord with boiled thread dipped in iodine, about half of an inch from the abdomen, and with sterilized scissors cut the cord just below the ligature. Warm the whelp in front of the fire and rub it vigorously with cotton wool, taking particular care to wipe the nostrils dry. Afterwards place it to a nipple. It will probably give its first cry and this usually induces the bitch to begin licking and tending it.

(4) *The Breech Presentation* (See Fig. 4). This is a complication which often occurs, probably averaging one case per six puppies born. It can gravely endanger the life of the whelp, the bitch and the whelps which are to follow, unless dealt with promptly, or if the bitch is left unattended for some hours. Fortunately it is a complication which can often be dealt with without the veterinary surgeon's aid, and in view of its frequency every breeder should know how to act. The condition is one in which the hind feet are presented first, instead of the head, with sometimes only one hind foot emerging while the other is caught up in the vaginal wall. The chief danger lies in the fact that a really good whelping bitch will often so exhaust herself by her abortive efforts to expel it that the uterine muscles lose tone and she loses the power to pass the remaining whelps. Often during the delay

another whelp will pass down the uterine horn and into the vaginal passage, so creating the complication of two whelps presenting at the same time and becoming entangled. This then becomes a major complication. A

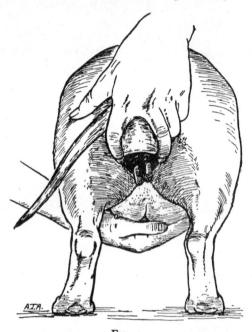

FIG. 4

The appearance of a breech presentation. Both feet have been presented or worked out with the fingers

bitch will frequently expel a breech presentation whelp unaided. She should be allowed reasonable time to do so. If she fails you must assist her.

Action. Get the bitch on a table. If you have an assistant let him hold the bitch's head firmly under his arm, place the right hand under the abdomen and with gentle pressure raise it. At the same time the assistant with his left fingers can stretch apart the vulva. With a piece of cotton wool (to counteract "slippiness") grasp with your left hand the foot or feet which are visible.

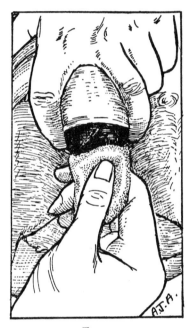

FIG. 5
The breech being delivered. First stage. The feet are now grasped firmly in cotton wool

DO NOT LET GO OF THEM (see Fig. 5), otherwise the bitch will withdraw them and may make no further effort, and your opportunity is lost. With your other hand pass a forefinger up the vagina, feel for the other foot, straighten it out, work it out, and grasp both with cotton wool. Again insert the forefinger of right hand, wait for a labour strain from the bitch and assist by working the whelp out a little further, at the same time exerting firm but gentle pulling pressure on the legs with the left hand. You will find, when the legs are entirely clear, that you can work about an inch of the whelp's rear end out. Now, still using cotton wool, firmly grasp in your right hand what is out, raise the uterus with your left, and in a DOWNWARD direction (NOT HORIZONTAL) PULL, and maintain a firm, determined tug (not a snatch). Remember, the whelp has GOT TO COME; don't be afraid of damaging it or the bitch—nine times out of ten you won't. PULL UNTIL IT IS OUT, steadily but not violently (see Fig. 6).

Note that emphasis is laid on the necessity for minimum delay in a breech delivery. If the envelope

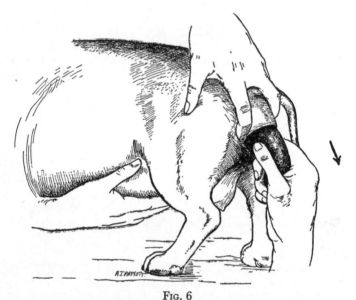

FIG. 6

The final delivery of a breech whelp. Arrow indicates downward and outward pull

covering the whelp happens to be torn and the whelp's head is uncovered it can't breathe and after a few minutes in this condition it will die. Or fluid may enter its nostrils and it is "drowned". It is better to get the "breech puppy" away, dead or alive, in order to save the whelps to come, and possibly the bitch. If prolonged, the whelp may be very weak and half dead; therefore sever the cord, clear the membrane, open the mouth with your little finger, draw forward the tongue, blow or breathe into the mouth. Swinging the whelp up and down is also very useful. When you hear the first whimper you will enjoy the thrill of achievement, of having successfully delivered your first breech. This may sound difficult to the inexperienced. In practice it is well within the capability of the novice, and after one or two cases it becomes simple.

(5) An abnormally large whelp, stuck half way; the bitch won't try to expel it further and runs about with the whelp hanging. Again prompt action is required. It is similar in many respects to the breech delivery, except that the whelp being head first makes the task easier and there is less chance of the feet getting "snagged". Although the head is exposed, and you will have cleared the membrane, the whelp still cannot breathe, as its lungs are compressed by the restricted space; consequently speed, as in the case of the breech delivery, is equally necessary here.

Wait for the intermittent straining by the bitch and work your right finger round the whelp at each strain with a "drawing down" motion, pushing apart the lips of the vagina (vulva) with the fingers of the other hand. If unsuccessful after a quarter of an hour, S O S your veterinary surgeon.

(6) The after-birth not coming away with the whelp, and only a small portion of cord exposed.

Action. Sever what there is of the cord exposed, as explained at (3), and proceed similarly with the whelp. This condition does not call for urgent veterinary attention, but you should always tell your veterinary surgeon that the after-birth was not passed. At his routine visit after the whelping he will inject a drug which will cause the uterus to contract, and so expel the after-birth and evacuate from the uterus any membranes left behind.

(7) A bitch is sometimes too fat or too heavy in whelp, and when a pup has arrived she cannot reach her end part to perform her immediate duties. Here you must do them for her as already described.

CHAPTER XI

THE last chapter described certain complications which can be successfully dealt with without the aid of a veterinary surgeon, if one is not available.

There are, however, more severe types, treatment for which is quite beyond the capabilities of anyone other than a skilled surgeon. The reader need not be alarmed or imagine that any of them occur frequently, but a brief description is advisable as it should be known that they MAY happen. Such knowledge should result in intelligent co-operation with the veterinary surgeon, the theme of Part II of this book.

A major complication is always a possibility if nothing has happened after three hours of real labour "heaves". If veterinary assistance is likely to be long delayed, or unobtainable, assistance may be attempted by the owner by inserting a finger right up the vagina and into the pelvis and endeavouring to twist, turn and work the whelp over the brim of the pelvis, or clear whatever obstruction is there. Complete sterilization and scrupulous cleanliness is vitally necessary whenever a novice makes a digital exploration of the vaginal passage. If subsequently a Caesarean section has to be performed dirty manipulation would cause a septic condition which would seriously affect the post-operative result. Therefore, wash the bitch's hind parts with soap and water, and whenever the fingers are inserted into the vagina the hands should first have been thoroughly sterilized with Dettol and water and the fingers smeared with medicated vaseline.

EXAMPLES OF SOME TYPES OF MAJOR COMPLICATIONS.

(1) Lateral deviation of the head. Instead of the nose passing over the brim of the pelvis the head is turned sideways and the neck only is against the orifice. All labour movements of the bitch aggravate the condition. Here injections of ergot or pituitrin would be absolutely wrong, as contracting movements would tend to dislocate the neck and tire the bitch. The veterinary surgeon is frequently able to turn the head and deliver the whelp with forceps, but don't be "fed up" with him if he decides to crush the whelp and bring it away piecemeal. This one lost will save the others and the bitch.

(2) Two whelps presenting at the same time down each horn, meeting simultaneously at the exit of the pelvis and becoming "locked". Pretty grim, and a sound argument for warning a veterinary surgeon at the three-hour period, as previously emphasized.

(3) Three whelps presenting simultaneously and getting entangled. Grimmer still, and only an operation can relieve. The writer had this experience with a 6½-lb. Miniature. Only two had presented together at first, and while working on these a third presented itself to mess things up even more—a tangled mass of three whelps over the pelvic brim. The veterinary surgeon had never had this complication before. He could do nothing on the spot; it was a case of a shot of morphia, hot water bottles, a dash by car to his operating table, and operative removal. All the whelps were lost, the bitch was terribly ill for three days after the severe manipulations but pulled through, and afterwards was a first-prize winner again, and promoted from the Kennels to be one of the house dogs as a reward for her pluck. Had she been left unattended in an outhouse from the start of this whelping till morning, her suffering would have been terrible and she would have died.

(4) The puppy presents "upside down". This is a very difficult complication. Normally the back should be

uppermost with the head and rump turning downwards and towards each other, and in this position the whelp slides comfortably over the pelvic brim. With the complication of the "belly" uppermost, with either a head or a breech presentation, when the bitch strains the puppy is forced upwards against the roof of the pelvis. They can rarely be delivered without forceps and if delivery is prolonged piecemeal removal is advocated.

(5) The front legs and feet facing backward and tucked on the chest, instead of forward under the nose. This increases the girth or bulk of the whelp in the region of its chest and impedes smooth delivery. Treatment is as for minor complication number 5 until the veterinary surgeon arrives.

(6) A "breech" presentation when the feet, instead of presenting, are tucked under the belly and only the rump or tail appears at the vaginal exit. A rather nasty case, but it can be dealt with by drawing the feet to their correct position, then bringing out one foot and then the other and proceeding as for the simple, uncomplicated breech.

(7) A fortunately fairly rare condition is when the backbone lodges across the orifice of the pelvis, and digital examination contacts only the back. It is a worse condition than lateral deviation of the head, and veterinary skill deals with it in much the same way as for that complication.

It is hoped that advice given so far in this book has made out a case for the necessity of being with your bitch at whelping time, and has shown some of the calamities that can happen if you leave her untended. Not only is your presence humane; such attachment to your bitch will also prove profitable to your Kennel Account. An owner not prepared to give this care should not breed.

CHAPTER XII

Complications which may affect the bitch after whelping—Conditions which can receive home treatment—The importance of knowledge of temperature.

COMPLICATIONS AFTER WHELPING.

(1) *Mastitis* or Mammitis, meaning an inflammatory condition of the mammary glands, or breasts. This is by no means uncommon and is caused by the milk not being extracted from one or more glands. It may happen whether whelps are or are not left with the dam. In the former case the whelps appear to ignore one or more glands, sucking only from a few of the teats. The trouble may be detected by the bitch being restless and off her food, and by her high temperature. The glands and nipples affected are swollen and appear to contain hard lumps; they are hot, reddish-purple, distended and very tender. The condition can be simple or very severe, according to how soon it is discovered and the promptness with which it is treated. In either case it can be treated by the owner. A veterinary surgeon's advice is desirable, and his intervention is necessary in extreme cases where a general blood-poisoning may supervene, in which event the trouble can prove fatal. Such a result can be due only to lack of knowledge, lack of treatment and lack of care on the part of the owner. The vital feature of treatment is that the milk must be drawn off. This is done by immediately applying hot fomentations, or by continually bathing the glands with hot water; every half hour is not too often. When the gland has been softened by this treatment, gently massage the nipple, which can be squeezed firmly. If a little milk is expelled all will be well provided you continue to draw off as much as possible. When the

milk has started to flow a lusty, hungry whelp may be put to the nipple. This will result in the milk being more adequately extracted. Such treatment may have to be maintained for three or four days, but it must be persevered with. If after all efforts the gland still remains hard, and no milk is expelled, the next phase, a little more serious, may be expected. This is abscess formation with temperature up to 104 or 105 degrees. Hot fomentation is again the treatment, to induce the abscess to burst pending the veterinary surgeon's attendance to open it with a scalpel. On bursting, a mass of blood-stained pus will be discharged and with proper treatment all will be well once more—temperature drops, pain subsides and the bitch improves in general condition. The maternal instinct is so pronounced in some bitches that they will suffer the pain and treatment while still allowing the whelps to suck from the healthy glands.

An authentic experience, with the course of treatment described, will give an instructive example of the veterinary advice and the correct method of carrying out treatment. The case involved a Miniature bitch, weighing barely six pounds. She whelped two puppies seven days too soon. They were too immature to live and died eight hours after birth. Here it was reasonable to suppose that mastitis might occur. A day later another bad whelping case occurred with a six-and-a-half-years-old Standard Smooth Dachshund, which proved fatal after Caesarean section—this being only the second whelping fatality in an experience of over twenty years. As a result of the Standard's Caesarean two moribund orphan pups were brought back from the vet's and hand-reared for twenty-four hours. We then tried to see if the Miniature bitch would suckle the orphaned Standard whelps; such an arrangement would be good for the bitch, to avoid mastitis, and would enable the pups to have natural bitch's milk. She took to them instantly and for three days all went splendidly. We then observed mastitis symptoms and the treatment already described was carried out, but we could draw off no trace of milk from one extremely hard and tender gland. The temperature had risen to 105 degrees and an abscess had appeared. We now deemed it wise to seek immediate assistance from our veterinary surgeon, and under his supervision the following treatment was carried out.

With continued hot fomentations the abscess was induced to burst, but the remaining condition had to be treated. First the wound was kept scrupulously clean by bathing with hot water hourly. Every two hours the hole was irrigated with acriflavine, 1 in 1,000 solution, and after each "wet" treatment the wound was left covered with a thick dusting of M and B powder. Temperature dropped and general condition improved. After three days of the treatment, with the abscess cavity clearing up well, the mammary gland tissue was protruding from the hole. This was all removed with a sharp pair of scissors. The acriflavine and M and B powder treatment was continued for a week, the edges of the cavity gradually approximated and complete healing took place. When it is realized that at the worst period the abscess sac was big enough to take a bantam's egg, was raw, purulent and the sloughed gland was exfoliating from it, it will be understood that mastitis can develop into a serious condition. However, if at the more serious stage the veterinary surgeon takes over and his advice is carefully followed no alarm need be felt. Provided one knows the correct procedure and treatment from early recognition to ultimate cure, the joy of success is ample reward for the time and patience bestowed. This instance is the most impressive in my experience of the "guts" and toughness which some maternal bitches will display when rearing whelps. With something so tiny fostering two Standard pups, at six weeks old nearly as big as their foster-mother, who for about a week was desperately ill and in intense pain, with the lusty pups sucking all the time close to a gaping, painful wound, and proving a zealous and jealous mother, we have an astonishing example of canine "grit" of the highest type. This bitch never looked back and is as fit and sound (though minus one mammary gland) as one could desire. As a reward she became a favourite house pet, instead of a kennel inmate. The orphaned whelps of the other bitch thrived beautifully, and one became the winner of a C.C. and two Reserve C.C.s.

(2) *Metritis*, or inflammation of the uterus, is not an uncommon complication of parturition. It can be caused by chill or exposure to cold after whelping; by foetal membranes, or after-birth, being retained, or by the unsuspected presence of a dead whelp. It may result from a very difficult labour, unsterilized fingers or instruments, or rough interference. The symptoms are lack of appetite, dullness, a marked rise in temperature, and swollen, discharging genital parts. Breathing is quick and irregular. In very acute cases collapse, coma and death ensue. Before such a result, treatment should be directed towards ridding the womb of its inflammatory and septic contents, but the treatment must be left entirely in the hands of a veterinary surgeon. Since the advent of penicillin injections the mortality rate has been vastly reduced. All the novice can or should do is to have enough knowledge to diagnose the condition. The symptoms described will help him in this respect. He may administer a laxative, as the bowels must be kept loose, and keep the patient very warm until the veterinary surgeon takes over.

(3) *Parturient, or Milk, Fever* is fortunately very rare. Another name is Eclampsia. The condition resembles a fit, alternating between spasmodic movements and a general paralysed or tetanized state. If such symptoms ever occur summon your vet immediately.

(4) *Early cessation of milk supply*. Detect in good time and hand-feed. The whelps will give you the warning

by crying continuously. Contented whelps rarely cry. Wean the whelps much earlier than under normal conditions.

(5) *Pneumonia.* After very difficult labour with complications necessitating extensive use of instruments, or administration of a general anaesthetic, pneumonia sometimes supervenes. Symptoms are high temperature and painful cough. The bitch will not lie down on her side, but sits up with drooping head. The eyes are red, the tongue dry and hot, the pulse rapid, and there is a catarrhal nasal discharge. It is a serious condition calling for veterinary treatment. The owner must keep the patient clean and very warm. Packing the dog's chest and body with Thermogene wool under a woollen coat gives her a lot of ease. After recovery, the wool should not be removed all at once but very gradually, a little each day.

TEMPERATURE.

Every novice may not know that a dog's temperature is not the same as that of a human. The normal canine temperature is 101·5 degrees, or nearly three degrees higher than in man. Thus a temperature of 102·5 degrees in a dog one would regard as slight, 104 degrees as high, and 105 to 106 degrees as very high to dangerous. As dogs cannot tell us when they are ill, the thermometer should be considered a most important piece of equipment which no Kennel should be without. It is our chief aid to diagnosis. Without it we should be quite unaware of the necessity of calling in a veterinary surgeon, if no other symptoms are apparent. So if one has a dog looking a little off colour and there are no symptoms at all, take the temperature as a routine measure. If the temperature is 102·5 to 103 seek veterinary advice at once. The ordinary clinical half-minute thermometer is used. It is first soaped or greased, then inserted into the rectum for about one inch. No force is needed, and dogs seldom resist, but the thermometer must be held in place firmly for from half to one minute to prevent it dropping out and breaking.

In cases of a nervous dog who does resist, control is gained by holding its tail firmly, then getting the animal between your arm and your body to steady it, holding its back legs with your left hand at the same time.

CHAPTER XIII

Dealing with newly born puppies—Complications which may affect
pups after whelping—Hernia, its avoidance and home-treatment
—Revival of a moribund whelp—Rearing orphaned whelps—
How to introduce a foster-mother.

Revival of apparently dead whelps. An early experience
may make so strong an impression that it remains to
influence one's subsequent actions. An incident is clearly
recalled that made the writer realize that though a
whelp may appear to be dead it can be resurrected, and
taught him never to give up hope until all possible
remedies have been tried. It happened in this way.

Circumstances had prevented the writer, then a novice,
from being present when one of his first "Scottie" bitches
was whelping. When he could attend her the bitch was in
great distress and almost unconscious. Two puppies had
been born three hours previously, but the bitch had
buried them in the sacking and they were dead. They
looked so dead that he wrapped them in paper intending
later to place them in the incinerator. On arrival of the
veterinary surgeon the dead pups were mentioned and he
asked to see them. The conversation took place in front
of the kitchen fire and, while one whelp was being held
up for the vet to examine it gave an almost imperceptible
shudder, and there was the faintest movement of a claw.
This was due to the combination of warmth from a hand
and from the fire. The vet said: "You've got a live pup
there—put it in the oven." Cotton wool was quickly put
in a cardboard box, both pups put in, and it was placed
in an oven probably hot enough to cook a meal. After
only a minute or two movements were seen from both
pups; later they were actually wriggling. In the mean-
time the vet was dealing with the bitch (a "piecemeal"

removal of the offending whelp with the complication of lateral deviation of the head), while the writer alternately rubbed the pups and held them close to the fire. The bitch fairly soon regained strength. Her intense pain having been relieved, she took an interest in the pups, licked them, and within two hours they had enough strength to attach themselves to a nipple. Both lived and one was ultimately shown.

After that the writer has never "thought" a whelp dead until he knew it to be "very dead".

The cause in the case referred to was quite simple—lack of attention from bitch or human, and exposure to cold.

With any of the major or minor complications already described, as for example a protracted and delayed breech, or where there has been undue compression on the lungs, a whelp when it is ultimately delivered will look lifeless. They are usually cold, long and drawn out, thin and flat, and appear so lifeless that they apparently have about as much chance of survival as the proverbial "snowball in hell". Many such cases can be "brought to life", though the inexperienced might reasonably have considered the whelps dead.

The following routine methods can be tried before the whelp is relegated to the dustbin. The thrill of seeing a movement or hearing a faint squeak in successful cases repays one for the trouble taken; it is a joy which must be experienced to be appreciated.

(1) Open the mouth and insert the little finger, very gently depress the tongue, blow down the mouth, or, holding the mouth close to one's own, breathe steadily in and out several times. With a pair of tweezers try to draw the tongue backwards and forwards, but no violence must be used, the movement being gentle and regular to correspond to normal inhaling and exhaling.

(2) Into a bowl place some very hot water (obviously not so hot as to scald the whelp) and immerse the whelp right up to its neck for a minute or two.

(3) For at least five minutes vigorously massage over the heart and lung area, and rub quickly to induce friction on the whole of the whelp's body.

(4) Hold the whelp very close to a hot fire, turning it round so that the heat penetrates all parts of the body.

(5) Hold the whelp's head in your right hand and the body in your left. Raise it to the height of your shoulder, then with a gentle swinging motion bring it down to the level of your knee. A violent movement is not necessary. This will often cause the heart to start beating.

(6) Drop a small quantity of neat brandy on the tongue.

If as a result of any or all of these measures you see signs of life continue your efforts with renewed energy; if life is present you can maintain it.

To sum up, many of the apparently "born dead" pups one hears of so frequently can be converted into "live pups" (and future C.C. winners) if some of the steps in-cated are taken promptly.

Hernia, its avoidance and treatment. How to avoid it. The subject of umbilical hernia could have been included in a later chapter, among the complications affecting the whelp after birth, but since one of its main causes can occur actually at whelping it will be more fitting to describe it and its treatment in a separate section at this stage. Its definition was given in Chapter X, where it was shown that anyone tending the bitch might cause a hernia through faulty manipulation, and the correct method was described. The second cause of umbilical hernia at whelping time is similar, but caused by the bitch. Some bitches, through fright or over-zealousness, attack the cord almost ferociously, and with one paw on the whelp will snatch and tear at the cord, and by tugging instead of quickly biting will draw the guts from the whelp's abdomen. If you observe this, intervene at once, soothe the bitch, and sever the cord for her in the correct manner already detailed.

How to treat it. The following method has been practised successfully:

(1) Cut from a cork a piece the size of a halfpenny and $\frac{1}{4}$-in. thick. (2) Gum this to a halfpenny. (3) Fix the halfpenny to a piece of adhesive tape, 1 in. wide and about 2 in. from one end. (4) Place a very small piece of cotton wool over the hernia, apply the cork over the wool, and fix the shorter end of the tape. (5) Take the tape once round the belly from left to right, once round from right to left (making an X) and finally straight round. (6) The tape must not be tight enough to restrict the abdomen at all, but sufficiently firm to put pressure on the cork to be transmitted to the hernia. (7) Make three or four turns with a 3-in.-wide bandage over the tape and stitch the ends. (8) If the bitch is not over-fussy she will probably not try to unravel the bandages. If she does, or if the whelp or other whelps tear at it, it may be necessary to feed the whelp by the bitch (under super-vision) every three hours, and isolate it at the intervals, attaching a wide cardboard circle collar round its neck to prevent it reaching the bandage. (9) If the cork is kept in position only forty-eight hours there will be considerable correction of the hernia. If maintained for a week a permanent cure may result. The reason for this is that the protruding gut is enlarging the traumatic opening in the weakened abdominal wall; if the protrusion is eliminated the opening automatically and rapidly draws together and disappears. (10) The same procedure can be followed for older whelps up to three or four months.

Many small or mild umbilical herniae in very young

whelps will disappear naturally without treatment. A more severe hernia in older whelps will produce a most unsightly "bump" later. Here veterinary advice should be sought; the apparent hernia may only be fat, but if it actually is protruding gut, surgical operation is usually necessary.

Prolapse of Anus. This is an eversion, or turning outwards, of the lower part of the rectum, and is caused in very young whelps by straining to pass faeces. Its appearance is that of an enlarged, very red fleshy mass from $\frac{1}{4}$ in. to $\frac{1}{2}$ in. protruding outside. This must be immediately reduced by pressing it back again with wet cotton wool, repeating the pressure every time there is a recurrence. Give the whelp plenty of liquid paraffin to drink. This loosens the bowels, reduces the necessity for straining, and as a lubricant eases the pain on passing faeces over the tender organ. Keep the part clean and moist by applying glycerine. Prevent the whelp dragging the everted rectum along the dirty floor. If the condition cannot be reduced thus simply, and great swelling, congestion and extreme tenderness persist, seek veterinary advice, when narcotics may be given, a local anaesthetic administered and astringent lotions applied. In extreme cases amputation may be necessary, but there are other lines of treatment the veterinary surgeon can adopt. His advice should be sought in all cases, but the above first aid home treatment is safe until the case is in his hands.

Milk Rash. For some reason the milk may not be suiting the whelp. The hair comes away in patches, and dry, flaky scabs appear and scale off. Treatment: give the bitch more vegetables, and orange juice, to reduce acidity of milk, wean the pups as early as possible, and treat the local patches alternately with T.C.P. and olive oil. Not a serious condition and clears up normally.

Eye Trouble. Failure to open at normal time, running or "watery" eyes, conjunctivitis, matter formation and tendency for lids to stick together, pus formation with a large bulge under the lid. In any of the above conditions simple first aid is to bathe regularly with a warm boracic lotion, applying it three times a day. "Golden Eye" ointment, or in cases of suppuration penicillin eye ointment, may be used, but only on a vet's prescription, since simple first aid should be adopted only until the veterinary surgeon has diagnosed the condition.

Intussusception. This is invagination of the bowel, or more simply a "telescoping" of one part of the bowel into another. It is not uncommon in young whelps up to eight weeks old, and is often present at the same time as prolapse ani. Symptoms are lack of appetite, marked abdominal distension, vomiting, rise in temperature, bloodstained and jelly-like discharge, crying out in pain, and marked tenderness on palpating the abdomen. The whelp is gravely ill, and no amateur treatment can be considered. The condition is nearly always fatal, though a clever veterinary surgeon might cure it by operation. His chances of success are increased if the owner has enough knowledge to recognize the condition EARLY.

Deformities. Hare lip, deficiency of toes, obvious malformation of joints or limbs, the latter being unequal in size, either elongated or shortened. In any such case accept your disappointment, cut your losses, and in the most humane way put the animal to sleep or, better still, get your vet to do it.

Inability to suckle. This may arise from a variety of causes, but whatever the cause carry on as for "orphaned whelp", described below, until normal suckling can be induced.

REARING ORPHANED WHELPS.

Briefly, the essentials are time, patience and perseverance. A whelp must receive some nourishment within three hours of birth. Any of the dried-milk preparations or baby foods may be used, and goat's milk is better than cow's. The writer prefers the fountain-pen-filler method, about one inch of milk per feed (for small breeds, and increasing proportionately with larger breeds) warmed to blood temperature and dropped spot by spot on to the back of the tongue, very gradually. This must be done every two hours day and night for the first week, and every three hours for the second. It is as well to have a helper to work in shifts. During the first two or three days, to give the whelp the instinct to suck later, place a piece of rubber dipped in the milk in its mouth about the size of a nipple and draw this backwards and forwards. When the whelps show intelligent signs of attempting to suck you can then change over to the other end of the filler, piercing the rubber several times with a broad needle. Its rubber end full will usually suffice for one feed for the smaller breeds. Hold your finger over the tip of the glass end so that pressure has to be exerted by the whelp's tongue to expel the milk from the rubber end down the throat. This also ensures that the feed is taken slowly. Once you can get whelps to suck in this way you can rear them.

Methods used by other breeders are to make the whelps suckle from a doll's feeding bottle, or a premature baby's feeding bottle, according to the size of the breed; or for very small whelps of any breed a paint brush with about one inch of hair well soaked in milk will encourage the whelps to suck and taste milk for the first few days.

Cleanliness and sterilization with any method is vital.

The addition of glucose to the warm milk feed is advantageous.

Amounts depend entirely on the size of the breed and the requirements of each whelp, which often differ. Generally they know themselves how much they want, and you can cease feeding when they lose interest. Never force more down, because you think the whelp should have more, if it has taken a reasonable amount and then desists. Over-feeding causes distension. During the second week the period between feeds may be increased by one hour, and after the second week a lapse of six to eight hours during the night without a feed is permissible providing they do not cry, which would indicate hunger. Diluted T.C.P. is advised for wiping the whelps clean after each meal. During the third week an alternative feed may be a thin but nourishing gravy or broth *via* the bottle, or spoon if lapping has been learned. A little later a small amount of raw egg beaten with the milk is strengthening. Regularity in feeding times is essential. Provided the whelps have taken sufficient nourishment by these methods for three weeks future feeding is identical with that for ordinary weaning of whelps which have been suckled from birth. At three weeks, scraped lean raw meat rolled into a ball the size of a finger-nail is sound for most breeds. It corresponds with the cutting and appearance of teeth, indicating that a more solid diet can now be introduced. Give one feed a day of orange juice, and also once a day drop a few spots of liquid paraffin on the tongue. An occasional fragment of Brand's Essence (about the size of a pea) placed on the back of the tongue is a good stimulant. A whelp reared in this way will usually learn to lap from a spoon much earlier than one fed by the bitch normally.

Whelps must always be kept very warm in a heated room, and have a hot-water bottle well covered with flannel under their blanket.

An important fact which is often not known is that, just as it is vital to get food into the whelp, it is important that waste products be removed. The bitch normally briskly licks the whelp's parts at, or after, a feed and the friction and warmth of the tongue induce urination and defaecation. This must be imitated and one way is after each feed to stroke the parts with cotton wool or a piece of flannel, either dry or dipped into hot water. Continue for a few minutes and you will find that functioning takes place. Without regular functioning digestive and other bowel disorders will occur, which will weaken the whelp and lessen its chance of survival.

Foster-mother. Although many whelps have been reared by hand from birth in some such way as just outlined, the introduction of a foster-mother and her management must be described, as it will save much time and labour, and bitch's milk is better than any substitute. Any breed about the same size as the normal dam can be chosen. If she comes with one or two whelps of her own she may prove chary of the strangers. This is overcome by keeping her away from the whelps she has to adopt until the following has been done: if she has any sacking or bedding of her own rub the whelps all over with this; put her whelps with the new ones and rub them together; go to the teats, squeeze out some of her milk and smear this all over the orphans. All this is done with the object of making the whelps new to her smell like her own. This done, she will probably settle down at once and let the orphans suckle from her. If the foster-mother has no whelps of her own smear with her milk those she has to adopt before introducing them.

Cats will often make excellent foster-mothers.

To conclude this section a statement made earlier is repeated—namely that although some rather grim details connected with whelping have had to be given it must be remembered that in the majority of cases the process is an interesting and happy event for the bitch and her owner. As this is not always so, however, every breeder should know what may occur and know how to act if skilled assistance is not immediately available.

CHAPTER XIV

An example of a difficult whelping with complications—The value of penicillin—Examples of inexperience.

A complicated whelping

THE reader has now been taken through phases of whelpings, beginning with general principles, passing on to detailed explanation of normal whelpings and finally through a summary of major and minor complications.

The account of an actual whelping case should be of interest, also of value in impressing on the reader the fact that whelping may not always be just the simple biological function he may have imagined it to be.

It would be unwise to confine a book of this kind only to breeds of which the author has had personal experience, so the experience of an authority in another breed has been included.

This case is given pride of place because it is really an amazing one. Its exceptional nature will be obvious to the reader in the light of the knowledge he has now gained.

I am indebted to an eminent breeder of Boxers for permission to tell this story of one of her champion bitches.

The bitch went through the weeks of her pregnancy in a perfectly normal manner; her appetite had been good throughout and she had exercised normally and alone. She had carried her puppies well without any sign of discomfort and had whelped normally, five puppies being born without any great effort, and the bitch appeared to have settled down contentedly in the customary way. The veterinary surgeon thought otherwise, however—and how sound it was that he had wisely been called in for the routine post-whelping examination! He considered that there were still two unborn whelps which were moving down only very slowly. An unbroken vigil was kept during the night but the bitch was comfortable and seemed quite content with the family she had already produced. Temperature was normal, but the veterinary surgeon's examination revealed a lump which could have been a knotted muscle, but which he distrusted. *Five* days later there was a rise in temperature and it was decided to make an X-ray examination, a very sound and necessary course. It revealed nothing. The owner's hopes rose but were dashed when the veterinary surgeon stuck to his original diagnosis, maintaining that the shadow of the ribs could easily be concealing something. Caesarean section was considered, but before resorting to that operation a further effort was made with warm douches, and by further injections of pituitrin to contract the uterus and induce delivery. Penicillin injections had been given every day after the birth of the first five whelps. (Note carefully this fact.) During the fifth night the bitch, being particularly clean in her habits and using only a certain field for her purpose, made several excursions to her particular spot in that field. Next morning, temperature was normal, discharge recurred, and the unusual lump had gone. While the owner and veterinary surgeon were puzzling over the situation another of the owner's bitches appeared on the scene, looking worried and not quite certain whether she was doing the right thing, but obviously feeling that something needed attention. She was carrying in her mouth a dead puppy! One would have thought that this would end the owner's troubles, but the veterinary surgeon was still not satisfied; he did not consider that the uterus was yet quite normal, or completely emptied. Two more days passed with no untoward event, except that the bitch's milk supply failed and the pups had to be hand-fed. The bitch was grateful for this assistance, her chief anxiety being that only one whelp was fed at a time. She moaned quietly to herself about those not yet fed but was quite happy again when the whole family's meal was completed.

On the *seventh* day, and three days after the field birth, temperature rose again and excessive discharge appeared. Pituitrin was again administered, but this time a close watch was kept on the bitch. She was confined to a room and allowed no further visits to her field. During this night the bitch was very sick, and quite suddenly there was an upheaval and a most hideously deformed puppy was delivered, bent over backwards with head and hind feet entangled. It was dead. For a week afterwards there was a continual and very severe haemorrhage, the bitch becoming extremely ill, and only the care of a very clever veterinary surgeon pulled her through.

When the haemorrhage ceased and the discharge had cleared and several days had passed normally, no one could fail to decide that this whelping had, despite its vicissitudes, been satisfactorily concluded. This, however, was not so. A fortnight after the end of this haemorrhage it was seen that tiny bones were being evacuated *via* the vagina. A possibility was that they were from the last deformed puppy, but as so many gradually appeared, and were from all parts of a skeleton, the veterinary surgeon was non-committal. Apparently he thought the whelping was still not yet through.

In actual fact, exactly *four and a half weeks* after the original birth of five puppies a complete leg and an almost unrecognizable head arrived. Pituitrin was again given and a little crushed torso came away; and further anatomical bony parts had to come before all could be assembled into an entire foetus. Not until after seven weeks from the first birth could the owner feel that her worries were really over and that the bitch would survive. The opinion of the veterinary surgeon was that this last puppy had died at about the sixth week of pregnancy, and was the cause of all the trouble. (A condition sufficiently septic to disturb the entire balance of the whelping phenomenon would undoubtedly have been set up.) The surviving puppies proved to be strong and lively, and full of fun and vigour.

The owner very naturally confessed to a thrill of pride in these hand-raised babies and was only too happy to say that the bitch was now looking quite well. In her own words, she regarded her as a "living miracle". She was also kind enough to recommend the special diet she had given to the puppies for the first three days after birth. It was: cow's milk and glucodin, then a mixture of slippery elm and glucodin of creamy consistency, gradually thickening, and alternating the glucodin with honey. Added to each feeding bottle were a few drops of rose-hip syrup, a pinch of calcium phosphate and a few drops of cod liver oil. (Many of the constituents of this diet would counteract deficiencies or impurities in what little milk came from the bitch at the very early stage, and would maintain the good start the whelps had made.)

A comment on one word of the owner's narrative. She regards her bitch as "a living miracle", and rightly so. She omits, however—doubtless from modesty—to point out that the miracle was due to three factors. First, her knowledge, her unremitting care, and her affection, things for which she deserves the highest praise. Second, the skill of the clever, conscientious veterinary surgeon with whom she co-operated so efficiently. Third, one word —penicillin. Intelligent, knowledgeable owner; painstaking veterinary surgeon; penicillin; in these three complementary elements lies the moral.

How well the story bears out a rule which has been stressed in earlier pages—the absolute necessity of the veterinary surgeon's visit after whelping is *apparently* over. Had there not been one in this case the bitch would undoubtedly have died from a metritis and an acute general septic condition within a few days of the original birth, her milk would have been contaminated before she died, and in all probability the whelps would have been lost too. Instead, the veterinary surgeon attended and penicillin was injected. Dogs react very favourably to penicillin; it is amazing how extreme septic conditions can be controlled and cured by its administration. For four and a half weeks after the original birth there was a putrid, disintegrating, decomposing whelp in the womb, and it was penicillin that prevented a general septicaemia. In the days before penicillin there would not have been a chance of the bitch surviving her gruesome ordeal; one can safely say that the case would have proved fatal within seven days. It is not permissible for an owner to acquire and administer penicillin himself but it is his bounden duty during many aspects of whelping to co-operate with the veterinary surgeon, who can use it in good time.

The value of penicillin

It is ironical that as a result of wars surgery has made enormous strides forward. The advances were incalculable after the 1914–18 war; to the present generation these have become routine and commonplace. Among the many innovations in medical, surgical and veterinary practice resulting from the last world war penicillin is one of the most striking. In the early days of the war it was in only experimental stages but now it is established as perhaps the biggest boon to mankind and animals by the power and rapidity of its control of sepsis.

In order further to impress on amateurs the advisability of seeking veterinary aid without delay, another example of the amazing efficiency of penicillin will be given, although it is not connected with whelping. The condition was that of Leptospira Jaundice, regarded by veterinary surgeons until quite recently as almost invariably fatal, and still regarded as a grave and dangerous disease. It is contracted from one source only, the urine of rats—particularly water rats, or rats who live near to a river or sluggish pond. Breeders whose

Kennels are close to rat haunts, such as farms, will know the condition and have had experience of leptospira. Others, and amateur pet breeders, will probably never have heard of it.

The symptoms are easily diagnosed if one has knowledge. A dog is perfectly fit, but an hour later has every appearance of being very ill. It lies down shivering, there is an astonishingly rapid loss of weight, half a pound to a pound in 2 or 3 hours, several pounds (according to breed) in 48 hours, and if the patient is still alive it is a skeleton in 4 days. There is an immediate drop in temperature to about 99° on the first day, down to as low as 96° or 95° on the following days. Later symptoms which clearly indicate the condition in its advanced stages are yellow eyes (where the "white" is normally), bright yellow gums and extremely foetid breath. There is also frequent and excessive vomiting, and complete loss of appetite throughout, with revulsion at any attempt at forcible feeding. Untreated, the disease is fatal within 4 or 5 days. Treated in time it can be cured, and a life saved with no unpleasant sequel. In the past four years the author has had three cases of leptospira—his Kennel adjoins a farm, and there is a slow-moving river within a quarter of a mile. He pays £25 p.a. to the British Ratin Company to keep the rats in check. By regular six-weekly visits they are kept completely under control, except during severe frosts and winter weather when they come up from the river for food, and are seen in the kennel enclosures and fowl houses. Each case occurred at winter time. The first was that of a dog who went "off colour" rather quickly. Beyond taking temperature, which did not show an increase, and not being unduly concerned at the vomiting—and ignorant of the symptoms— veterinary aid was not sought until towards the end of the second day. The case went four days and showed little response to treatment during the last two. By the fourth night there could not have been a thinner dog or one nearer death; the author would not have given one-tenth of a farthing for its chance, and expected to have to bury it at dawn. However, it had had a further large shot of penicillin by the veterinary surgeon last thing at night, and when visited again in the morning a crisis had been reached and passed; there was visible improvement, a tail feebly wagged and an eyelid faintly winked. The improvement was maintained, steadily increased, strength was gained and glucose and barley water were gratefully consumed.

Just as the onset of leptospira is dramatically sudden, so also amazingly rapid is the recovery; flesh and weight are rapidly regained and within three days of the crisis one has a beloved pet running around again.

Other forms of home treatment are absolute rest and quiet and utmost warmth, woollen jacket and hot-water bottles and maximum heat if kennels are electrically heated. Forcibly administer glucose and barley water, sponge the eyes and keep the hind parts clean. Scrupulously disinfect any spots where the patient has urinated —to avoid infection of other inmates. The veterinary surgeon will at once administer penicillin and liver extract, and all other treatment must be left to him— but penicillin is the real answer.

The dog referred to above subsequently gained one C.C. in this country and later became an Australian Champion.

The second case was that of a bitch a year later, at winter time. Only one day elapsed this time before the author "wondered" if the bitch's malaise "might" be leptospira, but wondering soon gave place to action, a veterinary surgeon's visit promptly resulting. Diagnosis was confirmed and a similar course of the disease was experienced, but though unpleasantly alarming it was not so severe. It was serious enough, however, for the author not to have "raised the kitty" on the chances of recovery by more than a full farthing! One day had been saved and one day's earlier treatment with penicillin had been given. The same dramatic recovery to full health resulted, and the bitch is still a leading Kennel member.

The third case happened in winter, too. But "you can't sting a nigger in the same place twice"; experience had taught us not to be caught napping again with leptospira. The case was a winning bitch puppy—incidentally, due to compete at the L.K.A. the next day, but things happen this way in dogdom and diehards philosophically accept such disappointments. At 10 a.m. she was running and playing normally. At 11.55 a.m. she was seen lying alone away from the other dogs, and shivering. There was no stopping to "wonder" if it "might" be leptospira! She was weighed and showed a loss of half a pound. Temperature was taken and found to be down to 100°. A veterinary surgeon was in attendance at 12.40. He suspected that the tentative diagnosis was correct, and penicillin was at once given. By night there was a further lowering of temperature and loss of weight, vomiting, and no food would be taken. The next morning the veterinary surgeon again came, confirmed that the illness was leptospira and a further concentration of penicillin was administered. It was naturally assumed that the bitch would be critically ill, as in the two previous cases. But during the day there was no further loss in weight or drop in temperature, and although no solid food was taken glucose and barley water and warm milk were swallowed keenly by the bitch of her own accord. When finally settled down for the night the patient was quite perky, but as this was only thirty-six hours from the onset, it was feared that by the next day she would be found to be very thin and weak, and with the yellow gums appearing. However, when inspected first thing the next morning, instead of being found lying semi-comatose she was walking about and wagging her tail and only a little shaky on the legs. More penicillin was given. By the evening the bitch looked better altogether. Temperature was 101°, virtually normal, and no further loss of weight beyond the original had occurred. She was offered meat, which she took hungrily. The next day, and only three clear days from the onset, she was to all intents and purposes quite well, whereas at this period she would have been approaching the worst stage of emaciation if the disease had taken its course. She was confined for three more days; she was eating well, functioning normally, with normal temperature, and seemed only to resent being kept away from her companions. At the end of seven days from the onset she had actually put on half a pound over her normal weight, due to extra feeding and no exercise.

These three cases have been described to emphasize the power of penicillin in the prevention and control of septic conditions if it is administered early enough. In the first case two days elapsed and in the second case one day elapsed before treatment with penicillin was begun. In the third only three hours elapsed between the first signs and the injection of penicillin, and as a result of thus nipping the disease in the bud the condition was actually neutralized. The lesson is obvious: early diagnosis, early treatment.

Elsewhere in this book reference has been made to the importance of temperature, and of the thermometer as an indispensable article of equipment. The foregoing is yet another example of its value as one of the main aids to early diagnosis of illness. In the case of leptospira, however, it is a rapid *drop* in temperature, not the more usual *rise*, which gives the clue.

There is an inoculation against leptospira jaundice which is inexpensive and claimed by veterinary surgeons to give complete and permanent immunity. Where rats are prevalent, or there is any risk of leptospira being contracted by contamination from their urine, Kennel owners are strongly advised to have every Kennel inmate vaccinated.

Leptospira is infectious or contagious only by contamination by the urine of a dog suffering from it. It is not
likely to cause an epidemic in a Kennel, provided all
other dogs are kept away from the actual patient. Consequently, a patient should be completely isolated and its
urine should be disinfected and efficiently removed.
Subsequently all its bedding should be burnt.

Examples of inexperience

We must now revert to whelping. A further example
is given with the object of showing a complete contrast with the first case mentioned so far as care and
treatment (or lack of them) are concerned, although the
conditions were similar. Comparison between the two
results will be inevitable.

A veterinary surgeon had been asked if he could recall
the longest period he had known of a bitch producing
whelps or foeti after the actual due date of parturition.
Since his experience of such a case had been recent he
could quite positively say the most protracted case in his
memory was three weeks and one day. A bitch was
brought to him critically ill, semi-conscious and unable
to stand, but very obviously due to whelp. To his horror
he received the following history. The puppies had
been due three weeks previously. The bitch had
seemed quite well, however, and there was no alarm
when she produced two live puppies a week later than
the due date. Nothing further was done. The puppies
died because the bitch "did not seem to be going
on very well", and it was eventually decided to
bring her to see him because "she seemed to be getting
worse".

On examining and palpating the uterus the veterinary surgeon came to the conclusion that there were
several whelps in the uterus, obviously dead. If he
did not operate the bitch would certainly die in a few
hours, whereas if he operated there was a remote chance
that she might live. Operation would involve complete

removal of all the organs. It was carried out. The womb contained a conglomerate mass of decomposed, dismembered whelps. The bitch rallied and lived for two days, and although the sepsis was controlled by the immediate injection of penicillin she was so weak that she succumbed to the post-operative shock.

The second object in describing this case is that it will serve as an indictment of cruelty against owners. Censure in this case, however, can be tempered by the fact that the cruelty was due entirely to ignorance, the owner being an amateur with a pet bitch and she thought she would "like it to have some puppies". Unfortunately, throughout the country there are many such pet owners with the same thought. Although their motive can be admired, their lack of intelligence or knowledge must be condemned.

Finally, as a contrast, the following has been chosen as the most humorous example of whelping inexperience.

A foreign lady had had a service to her bitch from one of our dogs; she was quite a novice. The author expected that she would be fussy when whelping time came, and a telephone call each evening of the week prior to the event rather proved it. The day the bitch was due to whelp the owner sounded more than unduly anxious, so it was decided to go over and assist her with the whelping. Arriving at 8 p.m. the usual imminent signs were evident but no actual labour. Thereafter till about 2 a.m. the night was punctuated by increasing jitteriness on the owner's part. In the bitch's interest it was tactfully suggested to the owner that she should go into another room and make herself some tea until there was anything to report. The bitch was restless, going to the door and scratching it, and as she had not been let out to relieve herself for some little time, the owner was asked to give her a walk in the garden for a few minutes. Shortly after an ear-splitting shriek was heard, loud enough to wake all the neighbours. The author and his wife hurried out and by torchlight perceived the owner at the extreme end of the garden, rooted to the ground. At her feet was the bitch with a whelp just being delivered! As this was not quite the place for the proceedings to continue they side-stepped back to the whelping room, the author carrying the bitch while facing his wife, who was supporting the whelp, the latter being delivered but the cord and placenta still attached to the bitch.

The routine procedure being followed, the whelping was concluded normally. However, the author firmly believes two things. Firstly, that the bitch, having been thoroughly bored with days of undue human fussing over her, had decided that when she was certain that she was going to have her whelps she would get as far away from humans as she could. Secondly, that had he not been present, the owner would have dialled 999 for the Fire Brigade, Police and Ambulance!

Another case of complete inexperience is the one of an owner ringing up her veterinary surgeon in a state of great agitation and informing him that her bitch was having puppies "*via* her mouth". The veterinary surgeon actually found it difficult to convince her that this could not possibly be the case. Only after he had explained that it was quite normal for a bitch to carry a pup about in her mouth, and resent anyone trying to take it from her, did it dawn on the owner that she had not beheld a whelping miracle. The owners were a married couple with two children, and one would have imagined that they would have had some knowledge of the facts of life—at least enough to know that there is marked similarity between the biological processes of humans and canines. One feels that this bitch might have had a rough time if the whelping had proved difficult . . . if the owners were concentrating attention on her mouth!

Another true example of inexperience—that of the lady who quite definitely thought that as her bitch had ten nipples it must invariably have ten pups at each litter, and was quite annoyed when it didn't! One wonders whether she would have expected always to have twins herself.

. . . . .

This book has been written for the amateur pet breeder every bit as much as for the novice breeder-exhibitor, who probably already has some knowledge of the subject. Owners of stud dogs in all breeds will undoubtedly have had experience of amateurs bringing pets to be mated, and will have gathered from questions asked that their visitors were ignorant of the fundamentals of the process to which they proposed subjecting their bitch. For many years now the author, after carrying out matings to bitches of amateur owners, has done one of three things: given them a half-hour talk on what to expect before, during and after whelping time, or given them written instructions on the subject, or lent them, or recommended to them, the breed book dealing with the elementaries of whelping. If other genuine dog lovers, including commercialized stud-fee earners, would realize how much suffering could be ended for in-whelp bitches, and to what extent the bitch mortality rate could be lowered, they, too, could give the amateur pet owner guidance, or at least advise him to read a book on whelping. Only a little knowledge, if based on principles, is preferable to the abysmal ignorance which the writer has met far too frequently. The "expectant" and her whelps would silently though most gratefully appreciate the extra attention which intelligent knowledge would substitute for cruel, though ignorant, neglect.

INDEX

159